ale[illegible]

E[illegible]re

stories

Alan Gravel

HAULIN' TRASH AND PASSIN' GAS

HAULIN' TRASH AND PASSIN' GAS

TACTICAL AIRLIFT AND AERIAL REFUELING IN VIETNAM

Alan Charles Gravel

Deeds Publishing | Athens

Published by Deeds Publishing in Athens, GA
www.deedspublishing.com

Printed in The United States of America

Cover design by Mark Babcock.

ISBN 978-1-950794-21-8

Books are available in quantity for promotional or premium use. For information, email info@deedspublishing.com.

First Edition, 2020

10 9 8 7 6 5 4 3 2 1

CONTENTS

This book is dedicated to:

Alan Watson Gravel

"Ali"

January 11, 1970 – April 19, 1980

A unique little guy who brightened our world while he was here

and

Sharon Lynne (Aasen) Gravel

"Sheri"

August 8, 1946 – March 1, 2016

She inspired and enabled the best in me

Painted Desert, 1971

INTRODUCTION

I began writing some of my stories from Vietnam just to be sure I did not forget them. I had no specific plan in mind about what I would do with them and was far from sure that anyone would be interested. Over the years, I had several occasions to use the stories with some accompanying photographs when interviewed by students from Georgia Tech and Oxford College of Emory University. Occasionally some event would trigger my memory and I would either write an additional story or two or at least add the story to my list that I might write someday. In 2019, the Atlanta Vietnam Veterans Business Association began to gather stories from its members that eventually were published as *I'm Ready to Talk*. In conversations with Bob Babcock of Deeds Publishing about that book, I began to think that all of my stories about Caribous and Tankers together might be interesting to a few people outside my immediate family. At the very least, I would have my most significant memories of Vietnam recorded together in one place.

It occurred to me recently that I am bringing this process to a conclusion 50 years after the events that I have written about. I am sure that there are some who would say "put that behind you and move on with your life." Actually, I did that many years ago and have not lived a life tortured by, or even dominated by my memories of Vietnam. However, the experience of participating in armed conflict was formative to me in a way that few other life experiences were. The lessons I learned in Vietnam about myself, life, human nature, and in a nutshell, duty, honor, and country have made me the person I am today. I am not willing to put those things behind me. Even if I chose to, I could not.

This is not a book of war stories about firefights, hand-to-hand combat, or the tragic loss of a best buddy. Those stories are told by heroic front-line soldiers who had very different experiences from mine. These are, quite simply, *my stories*. These are the things that happened to me. These stories tell how I experienced the Vietnam War. I do not tell them to illustrate courage, or daring, or heroism. They are just what happened. I hope the reader will discover in this book some elements of the Vietnam experience that will broaden their understanding of that War.

ABOUT THE TITLE

In C-7A Caribous, we frequently called ourselves "trash haulers" not as an insult to the passengers and cargo that we hauled around, but as a reference to the fact that we routinely hauled a wide variety of loads. In one day, we might carry military and civilian passengers, US and Vietnamese passengers, live animals, ammunition of all sorts, food, mail, lumber, beer, and a multitude of other cargo items.

In KC-135 tankers, we would often say we were going out to "pass gas." Of course, the fuel we off-loaded was not actually gasoline but JP-4 jet fuel. Nevertheless, "passing gas" had a ring to it that "aerial re-fueling" did not. Thus the title. Perhaps a bit inelegant and undignified, but appropriate in the context of a war zone where shorthand communication, acronyms, and slang were the norm.

GETTING READY

MY LIFE BEFORE

I was born in Alexandria, LA in 1945. I was the third of four children with an older sister, an older brother, and a younger brother. Our Dad was an accountant and ran his own small bookkeeping service. Mother had been to business school to learn typing and shorthand, but she was a stay-at-home mom.

Our Dad was diabetic so was never able to serve in the military. All four of his brothers served, one as an Army officer, one as an Army Air Corps/Air Force pilot and one as a Navy enlisted man. The fourth was an Army Air Corps Second Lieutenant who was killed in a training accident in a B-24D Liberator near Biggs Field at Fort Bliss in El Paso, TX in January, 1944. Mother's three brothers all served, all as enlisted men, one in the Marines and two in the Army Air Corps/Air Force. Despite all that, I did not grow up with a strong orientation to join the military. I don't know that I thought a lot about my future career, but what thinking I did, did not include the military.

Dad died at the age of 37 when I was 10 years old. He had been diagnosed with Type I diabetes when he was 16 years old and so was uninsurable. We grew up on Social Security Survivor Benefits and what Mother could earn, first as a school lunchroom worker and later as a secretary/bookkeeper for our church and then the Louisiana Baptist Convention. My Mother was determined that we were going to get a good education and learn to swim, the two things that she had never been given the opportunity to accomplish. We were very much involved in church and church activities. Aside from the spiritual aspect of that involvement, Mother felt that involvement in the church would provide us with opportunities that we would otherwise not have. She was right. We attended and worked in summer camps and attended retreats and other events that enriched our childhood.

In high school, being a late bloomer physically, and not being athletically talented, I played in the band. Bolton High School was an excellent school and had "accelerated" courses in math, science, and English. I was in all of them. I was on the debate team and we competed in state-wide contests, including the Glendy Burke debate tournament at Tulane University in New Orleans. We were not that successful, but the experience was enriching.

Much to the surprise of faculty, classmates, and even myself, I graduated with the highest academic average of the 316 students in the class of 1963. The school at that time did not designate a Valedictorian but chose another student who excelled in oratory to speak at the graduation. At the graduation ceremony, they recognized the top ten ranked students and there were some looks of puzzlement when I stood up. It was a City school and I lived out in the country. I was one of a handful of students who rode the bus to school. We were not wealthy or socially prominent. The odds-makers missed that one.

Throughout my senior year, I thought about where I would go to college and how I would afford it. I considered Georgia Tech that had a co-op program where you could work your way through. The guidance counselors encouraged me to apply for an academic scholarship to Louisiana Tech. My sister had attended there for one year before transferring to Blue Mountain College in Mississippi. My brother was there studying Forestry. It was known as a good Engineering school. When I was awarded the scholarship, the decision was made.

Because my older brother was involved in the Baptist Student Union at Tech, I immediately became involved as well, even attending the Pre-School retreat at the Caney Lake Camp near Minden, LA. There I met and became friends with several people who remain my friends today.

BSU was another activity that broadened my limited experience as I was able to serve in leadership positions and in Summer Missions programs the summers before my junior and senior years.

In 1965, I served as a summer missionary in a church in Bangor, Maine. Several members of the church were Air Force pilots who were involved in the B-52 "Chrome Dome" nuclear bomb armed flights toward Russia. I also served for about half that summer in Machias on the eastern Atlantic coast where I worked closely with an Air Force radar station operator who was a member of the church there.

In 1966, I served in Guyana, South America. We arrived there in early June after they had gotten their independence from Britain on June 1. I got an early taste of jungle bush flying when we flew with the Missionary Aviation Fellowship to the site of a British owned manganese mine at Matthew's Ridge to conduct a week-long Vacation Bible School. They stopped the mining trucks on the dirt road to allow us to land. Matthew's Ridge was not accessible by road, only by railroad that connected it with a seaport where the manganese ore was loaded onto ocean-going vessels. It was only a few miles from the site of the now infamous Jonestown. Years later, when I flew Caribous in Vietnam, I thought about that trip and how much my understanding of that environment had changed.

In my junior year, a new professor of Civil Engineering arrived at Tech. He had been an officer in the Naval Facilities Command for 30 years. After retiring from the Navy, he attended The University of Texas to earn a PhD. I was in his class, Hydrology, I think, in my junior year. One day I overslept and missed the class. I went to his office that afternoon to apologize and get the assignment for the next class. He told me, "Son, if you do not learn to get out of bed in the morning, you will never get very far in life." Excellent fatherly advice. I went that afternoon to buy an electric alarm clock that could not run down, no matter how many times I hit the snooze.

In the fall of my senior year, I had the same professor for another course, this time Ground Water Hydrology. About halfway through the semester, he stopped me after class one day and asked if I had any interest in going to graduate school. It seemed that everyone graduating in those days was going to Vietnam, so I told him yes. He assured me that he could get me a Public Health Service Fellowship. I applied, and with his recommendation, got the fellowship.

Sheri Aasen had come to Louisiana Tech as a sophomore when I was a junior. We had one date in the autumn of 1965 that did not go very well. Over the balance of that year, though, we saw each other every weekday at noon-day chapel at BSU and soon developed a friendly acquaintance. I remember thinking that she would one day make some lucky guy a great wife but that it apparently was not going to be me. By the end of my junior year, I had accepted the "fact" that I would not meet the person I would marry in college and would have to wait until the next phase of my life, work, graduate school, military, or whatever, to find someone to share my life.

In the summer of 1966, Sheri went to West Virginia as a summer missionary and I went to Guyana. The next fall, the BSU director asked me to go over to Northeast Louisiana State College in Monroe to give a talk to the BSU there about my summer in Guyana. He told me to take one of the other summer missionaries with me who had served here in the States. I reasoned that Sheri was so good-looking that all of the boys at Northeast would listen intently to what we said. I was right.

We drove over to Monroe in my car and on the way back we began to talk seriously. When we arrived at her dormitory at Tech, we sat in the car and talked until 11:00 pm when she had to go inside. From that point on, we saw a lot of each other.

A few weeks later, on a Wednesday afternoon, I was in my 14th floor dorm room polishing my shoes and looking out the window when the thought suddenly popped into my head, "You *will* marry Sheri Aasen." Not "maybe you should ask her," not "this could happen," but "You will."

I asked Sheri to marry me on December 4, 1966. She said yes, so our plans logically evolved. I would finish my senior year and go to Texas for the 1967-1968 academic year. She would stay at Tech that year to finish her degree and graduate. We would get married in May of 1968 and go to Texas for me to finish my Masters. Beyond that, who knows?

In order to graduate in 1967, I had to attend summer school to pick up a Physical Education Course that I had somehow missed. Luckily, Tech had a new swimming pool and I was able to enroll in a course that got me Senior Lifesaver and Water Safety Instructor Certifications. I took two other courses just for the fun of it, one being Experimental Design. My senior year roommate was in the same boat, so we continued to room together through the summer. Tech had a new IBM computer (their first) and he had access to it so we would go over there at night and make the thing work as hard as we could solving iterative problems like backwater curves and things like that. That summer, Sheri served as a summer missionary in Japan, so it was good that I had something to occupy my time.

Summer School ended, I got my BS in Civil Engineering, and Sheri returned from Japan. We had a few weeks to catch up before school started for her at Tech and for me at Texas.

UNIVERSITY OF TEXAS TO USAF

When I entered graduate school at the University of Texas in Austin the autumn of 1967, the Draft Boards would give you a 2S Student deferment for graduate school. During my first year at Texas they changed the policy to limit 2S deferments for graduate school to one year. Sheri and I got married at the end of May 1968, and we settled into an apartment at 201 Oldham in Austin (since then demolished for the LBJ library). I was enrolled for the summer session. In June, true to the new policy, I received a re-classification to 1A which made me draft eligible. I appealed the re-classification to my home Draft Board in Alexandria, LA, not expecting to be successful but just buying some time to figure out what to do. I still needed one full semester of course work in the autumn and then one semester enrolled in a thesis course to finish my Masters' degree in Environmental Health Engineering.

About 30 days after the 1A classification, I received

my new classification and I was 2S again. I was shocked and a little confused but decided not to look a gift horse in the mouth. We resumed our life as newly-weds in Austin, trying to live on my $292.50 / month fellowship from the Public Health Service, but very happy and enjoying life.

About 30 days after that, I received an unexpected re-classification to 1A. After years of thinking about this, I can only believe that the 2S given in response to my appeal was an error, possibly a simple clerical error on the part of my home Draft Board. For whatever reason it happened, that 2S was an extremely significant stroke of good fortune because it allowed me to start the entire appeals process over.

I had learned that you could appeal to the Draft Board where you were living rather than your home Board. Austin at that time was a hotbed of anti-war protests and anti-military sentiment. That, combined with the population of 30,000 students enrolled at the University, meant that the local Draft Board had a year-long backlog of appeals waiting to be heard. I appealed my 1A classification to the Austin Board and never heard from the Selective Service again.

At the end of the summer, we moved to an apartment at 3816 Speedway in Austin. I enrolled in the fall semester to complete the last of my course work. After some discussion with the USAF recruiter in Austin, I enlisted in the Air Force. They sent me out to Bergstrom

AFB to take a flight physical and gave me a battery of tests that evidently qualified me for pilot training. I got a delayed enlistment so that I could finish my course work at UT, complete my thesis, and fulfill all other requirements for my MS in Environmental Health Engineering. I finished proofreading my thesis offset printing plates around midnight on February 16, 1969 and went to bed. Sheri woke me up about 3:30 and drove me downtown to get on a bus to San Antonio to begin my USAF career.

OFFICER TRAINING IN SAN ANTONIO

After a long day at the Induction Facility in San Antonio, we were bused to the Lackland AFB Officer Training School on the Medina campus about 6 or 7 pm in the evening. There we were issued some uniform clothes and assigned to our barracks. Each flight of 15 trainees was assigned to a young Training Officer who would be our Mother and Father for our time at OTS. Our TO was Captain Lommatsch.

On the first night Captain Lommatsch told us, "I have had championship track teams, I have had championship flicker-ball teams, I have had championship just about everything except academics. If ya'll do well in academics and keep your noses clean, I will make this easy for you." I since have suspected that he had hand-picked us to be strong in academics. We finished first in the wing in academics as a flight and three of us finished very high as individuals. So, things did go well for us.

The first weekend we were confined to quarters. The second weekend we were confined to the Medina campus. The third weekend we were restricted to San Antonio. After that we were free to go wherever we chose from about noon on Saturday to late Sunday afternoon. Since Sheri was still living in our apartment in Austin, which was only 90 miles away, she would drive down on Saturday morning and we would return to our apartment in Austin. This respite from the yelling and abuse and stress that is a part of basic military training made my OTS experience much more bearable. I have come to think of it as "outpatient" officer training.

We had it as easy as it can be but Basic Training under any conditions is rigorous and a rude shock to a bunch of college boys who are not accustomed to the discipline, long hours, and regimentation. Each day started with, "The time is 0530, the UOD (uniform of the day) is …." You had about two minutes to get out of bed, dress in the correct uniform with the correct headgear, make your bed, and be at attention out in the hall. After a few days, we just slept on the floor and left the bed made-up 24 hours a day. It was not worth the trouble to sleep comfortably and then try to do the impossible getting the bed perfect every morning.

One of the uniform items that we had been issued was what they called a "wheel hat." It was the traditional military dress hat and the rim was fairly rigid. Throughout the day as we entered and exited buildings we had to put it on and take it off. In about the third week, I developed a swelling on my forehead that was very sore and red. When I went to sick call, they made an appointment for me with a plastic surgeon at Wilford Hall hospital on the main Lackland campus. Wilford Hall was to the Air Force what Walter Reed is to the Army. The doctor quickly identified my problem as a sebaceous cyst that had been ruptured by the wheel hat and was inflamed. He made a small incision to remove what remained of the cyst, stitched it up and put a bandage on it. He gave me what he called "flight cap orders", a small slip of paper similar to a prescription that basically said I could wear whatever cap was comfortable to me, regardless of what the UOD was. I was to keep the "flight cap orders" with me at all times in case I was questioned about why I had on the wrong hat.

Everywhere we went on the Medina campus, we marched as a unit. About 15 Officer Candidates were in each flight and that group virtually always traveled together. We marched to breakfast, lunch, and dinner together. We marched to all classes and physical training activities together. We marched to all meetings of the whole wing together. As we marched across campus, there were always a number of other flights on their way somewhere and their Training Officer was usually trailing along with them. These TO's loved to find an

Officer Candidate in another TO's flight out of uniform or, in my case, with the wrong headgear. They would order our flight to stop and then approach me and dress me down as "too stupid to get the right hat" or something similar and call me everything but a child of God. I would not interrupt them but let them go on and on until they asked for a response. Then I would pull out the "flight cap orders" and without saying anything, hand it to them.

The first few times this happened, the whole flight, including me, was *very* anxious about what the outcome would be. Each time, when the TO realized that he had made a fool of himself, he would "hrrumph" and order us to "carry on," or something like that, hoping no one else had seen what happened. As time went on, these encounters became very funny. The whole flight could see the guy coming from 30 yards away and we would start muttering between closed lips about what was getting ready to happen. It got so funny that it became difficult not to snicker, smirk, and even smile or giggle even before the confrontation began. I like to think my sebaceous cyst provided some valuable comic relief for our flight that helped us bear up under the rigor of the training.

Each Saturday morning, we would have a big inspection of our barracks room and our persons. Half of the Officer Candidates were six weeks ahead of the other half, so we had an upper class and a lower class.

The upper classmen conducted the inspection and they were all about trying to intimidate the lower class. They wore black leather boots with taps on the heel and toe. They would "stomp" into the building, "stomp" up the stairs, and "stomp" down the hall until they reached each room. We quickly learned to use Pledge spray furniture wax on the floor, particularly at the top of the stairs and outside each room. If the upper classmen were not careful when they executed a "right face" or a "halt" their feet would go out from under them and they would end up on the floor. That tended to dampen their intimidation tactics.

My brother Sam was two years ahead of me in school. He had completed Officer Training and Navigator School and had been sent to Da Nang to fly as the "guy in back" (weapons officer) in F-4's.

When the time came for me to graduate and receive my 2nd Lieutenant gold bars, Sam was just returning to the US after a year in Vietnam as a young 1st Lieutenant with a combat tour under his belt. He flew into Travis AFB on a "Freedom Bird," caught a hop to Kelly AFB in San Antonio, and somehow got over to the Lackland Medina campus the morning of our graduation. He was too late to join Sheri and my Mother for the graduation ceremony, but he got there in time to swear me in as a 2nd Lieutenant. He postponed his reunion with his new wife (they had married on R&R in Hawaii) to be there for my graduation. A real honor for me.

Officer Training Saturday morning inspection

My brother Sam (left) swears me in as a 2nd Lieutenant.

PILOT TRAINING IN DEL RIO

After OTS we went back to Austin, packed our belongings in a U-Haul trailer, and moved to Del Rio to join Undergraduate Pilot Training Class 70-07 at Laughlin AFB. Just about the time we arrived, construction of a new off-base apartment complex was completed. A large percentage of the married students in 70-07 rented these new apartments, so it was virtually "married housing" for our pilot training class. This simplified car-pooling and facilitated all sorts of social interaction.

The apartment complex management had worked out a deal with a local furniture store to rent the furniture needed for the apartments. Most of our friends and classmates took the deal and had fully furnished apartments on day one. I looked at the cost of renting the furniture and thought I could do better. I joined the credit union and borrowed an amount of money that gave me 12 monthly payments equal to the rent payment. We were only able to buy a bed, chest of drawers, sofa, and a couple of folding chairs. Through the year we added a few other things, but we mainly just got by on less. Our ditzy downstairs neighbor came up to see Sheri one day and immediately said, "Oh, Monte was right, ya'll *don't* have any furniture."

The first 30 days of training were in a T-41, which was the USAF's designation for a Cessna 172. The instructors were civilians and the purpose was widely thought to be just sorting out the students to see who had the aptitude for flying and who didn't. The incident that I remember most vividly occurred one morning when the fog was so thick it was barely possible to drive, much less fly. Training was cancelled and we were sent home with the admonition that nine months after similar days in the past there had been a surge in births at the base hospital.

After a month in T-41's, we spent five months in T-37's. Also built by Cessna, the T-37 was widely thought to be a machine built to convert jet fuel into noise. The student and instructor sat side by side so there was no hiding your mistakes and confusion. We were introduced to the simplicity of jet power and somewhat higher performance. We could fly higher, faster, and farther and got an introduction to aerobatics and instrument flying.

The final and best phase of our training was in the supersonic jet trainer, the T-38. This was and still is an outstanding airplane, fun to fly, and a serious step up in performance. Fully fueled it weighed 12,000 pounds and in afterburner had 12,000 pounds of thrust. Technically you could takeoff, point it straight up and accelerate while climbing vertically. It held the time-to-climb records for many years and was the aircraft used by the astronauts when commuting between Houston and Cape Kennedy.

The first phase in the T-38 was instrument training. We spent about a month sitting in the back seat under the "bag" flying strictly by instruments. The instructor would get the aircraft into the air and to the training area and then we would practice our maneuvers while relying solely on instruments. I did well on this phase of the training, which lasted about a month.

The next phase was "Contact." Basically, this was learning the stick and rudder skills to take off, land, and maneuver the aircraft. The student sat in the front seat and flew mostly by visual references.

One of the more challenging emergency procedures we trained for was a heavy-weight single engine landing. Immediately after takeoff, with a full load of fuel, you would pull up into the traffic pattern, the instructor would pull one of the throttles back to idle, and you would have to land using only one engine. One day during Contact Phase, I was sitting at number 1 for takeoff when one of my classmates and his instructor were practicing this procedure. I was looking at the checklist when in the corner of my eye I saw a quick white flash out beyond the end of the runway. As I looked up, a vertical tower of black smoke formed, topped by a mushroom. The tower closed the runway and directed all aircraft to hold their positions. The fire trucks came racing out, drove off the end of the runway, pushing down scrub trees and fences, crossed Hwy 90 and reached the crash site out in the desert. By the time they got there it was mostly over. This was the only fatal crash that my class had all year, but it was a terrible one and I was an eye witness.

Shortly after this accident, the Air Force stopped training for the heavy-weight single engine landing. They realized that they were killing more instructors and students practicing that maneuver than would ever be lost due to the actual occurrence of an engine failure just after takeoff.

When I had completed my "contact" training, I failed the check ride because of my poor performance on the heavy-weight single engine landing. I took remedial training and failed the second check ride. I think I failed four or five check rides and finally came to the "9999" ride which was *really* the final one. If I had failed that one, I would have been washed out of pilot training and likely sent to navigator school. The check pilot was the Chief of Standardization.

At the appointed time, I arrived at the Chief's office and reported, "Lieutenant Gravel reporting, sir."

He replied, "Gravel, huh. You're not by any chance related to Sam Gravel, are you?"

"Yes sir, he's my brother."

"Yeah. Sam and I put a couple of rockets through the side of the USS Boston one night."

He was Lt Colonel Mehaffey and he had flown F-4's with my Brother as his weapons officer out of Da Nang. They had been involved in a friendly fire incident involving the Navy. The Air Force was totally exonerated—the Navy was at fault, but no matter whose fault, it was an unfortunate incident.

I flew well that day and scored in the high 90's on the check ride. That whole incident helped me learn about myself... that although I am not a very physically talented person, with enough practice I can eventually master some physical skills. The things I have been competent at in life, swimming, tennis, ping pong, etc. have been activities that I had the opportunity to practice extensively.

Many years later it occurred to me that the fatal crash that I had witnessed might have had something to do with the difficulty I had with the heavy-weight single engine landing. I literally never made the connection until years after I left the Air Force. At the time I had written it off totally to my difficulties mastering the eye-hand coordination skills.

There wasn't a lot to do in Del Rio and we trained 12-14 hours a day. On several weekends we drove over to the Pecos River Canyon and explored some of the Native American caves in the canyon walls. I still have a few arrowheads that I found there. We probably violated all sorts of laws, but ignorance was bliss.

Sheri was pregnant the first half of pilot training and was due around Christmas day. Because of that we stayed in Del Rio for Christmas when most of the class went home. It was quiet but very nice. We had our little Christmas tree and I bought our first television so we could see the Christmas music programs and movies. Alan did not show up on time. By the end of the first week in January, Sheri was getting desperate. Just before the doctor was to induce labor, Alan finally decided to arrive and was born on January 11. Classmates Mike Stowall and Denny Healy went with me to bring Sheri and Alan home from the hospital. Denny insisted on carrying Alan. His wife Susie was pregnant, and he was so excited in anticipation of fatherhood. Denny was one of the top pilots in our class. He was killed a few years later in an F-4 accident in Spain.

On toward the end of T-38's, we flew some out-and-backs to other bases. The first ones were with instructors and later we were trusted to go out solo. When my solo out-and-back came up, I chose to fly to my hometown, Alexandria, LA, to England AFB. My Mother still lived there so I let her know and she came out to see me and have a sandwich at the flight line snack bar. That must have been a very strange experience for her, a mixture of pride and concern for my safety. On the way back to Del Rio after dark, my cabin air malfunctioned and I almost froze before I reached Laughlin. I made one of the worst landings of my T-38 career because my hands and arms were so cold and stiff. Fortunately, I was the only witness.

The balance of T-38 time went reasonably well for me. I excelled in the ground school, academic portion, and was always a little behind in the flying part. When we got to the end, I think I was about 42nd out of 60 in the class. First in the class got his choice of assignments and it went downhill from there. The lowest guys in the class got B-52's and since I was just above them, I got the aircraft that few in the Air Force knew anything about, the C-7A Caribou. Looking back, I consider that very good luck.

T-37 trainer.

Me and the T-38. We called these our "steely-eyed killer" pictures.

Sheri pinning on my Air Force wings.

L to R. Me, Sheri, Denny & Susie Healey, Sharon & Mike Stowell.

C-7A CARIBOU

ON THE WAY

CARIBOU TRAINING IN ABILENE

After pilot training we reported to Dyess AFB in Abilene, TX, for Caribou school. The new class of about 60 students met the first day for an orientation briefing. Most were like me, recent graduates of Undergraduate Pilot Training. A smaller number were experienced pilots being transferred to Caribous from other aircraft.

I have a clear memory of walking into the room filled with all these new people and within 30 seconds or so encountering David Mitchell and thinking, "I could become friends with this guy."

Coincidentally, we had both rented efficiency apartments which were part of the Starlight Motel. David and his wife really enjoyed being around Alan who was about six months old and the five of us frequently went out at night for ice cream. Our lives have been intermingled ever since and we are best friends.

Two students were assigned to each instructor. One

would fly the airplane and the other would stand between the seats observing. In addition to short field landings, we flew some cross-country and instrument training.

One of my instructors said something that stuck with me: "When you go to Vietnam, you will face all sorts of hazards and challenges—ground fire, mechanical problems, personal conflicts, unsecure landing sites, etc. The hazard that will be most likely to hurt you will be bad weather."

What I found to be true was that bad weather was often the last straw. While you were trying to deal with other difficulties, weather would often compound the problem and exacerbate the difficulty you were facing.

At Dyess, there were two different types of airplanes, C-7A Caribous and KC-135 Stratotankers. Caribou students and instructors practiced short-field landings on a dirt runway off to the side of the main concrete runway used by the tankers. All short-field landing practice had to be full-stop landings to be able to practice the reverse thrust and braking. Touch-and-go landings did not teach the complete process of bringing the airplane to a stop within a short distance.

After each short-field landing, Caribous had to taxi back around to the approach end to take off again. Since there were three or four aircraft practicing at one time, we spent a good bit of time on the ground waiting our turn to take back off. While there, we would occasionally see a KC-135 tanker taxi out, take the runway, and start his takeoff roll. When they applied takeoff power, the whole world seemed to vibrate with the rumble coming from the engines, but the tanker would only slowly roll off down the runway. They would roll and roll and roll until it seemed they would go out of sight and then finally stagger into the air as if only barely defying gravity.

More than once I said to myself and to others, "I hope I never have to strap my ass to one of those things." As was often the case in the Air Force, 18 months later, after my one-year tour in Vietnam in Caribous, I was doing exactly that.

SURVIVAL SCHOOL IN SPOKANE

After completing Caribou training, Sheri, Alan, and I traveled to my home in Alexandria, LA, to get them settled in for the year that I would be in Vietnam. We had bought a mobile home to set up on my Mother's property so they could live there until I returned from Vietnam. I remember us driving out to Esler Field the day I left for Survival School in Spokane. I could not promise Sheri that I would come home safe and she did not ask me to. Mother seemed to respect that this was a solemn moment for Sheri and me, so she occupied

Alan's attention as much as she could. I am certain that somber thoughts were swirling in her mind as well.

When we arrived in Spokane for training, David and I were back in the same group. Survival School involved some academics, training in how to land in a parachute, basics in physical survival, Escape and Evade training, and Prisoner of War Training. Toward the end, we spent three days out in the woods with 15 men sharing one live rabbit for food. At the end of that we were "captured" and placed into a POW camp where we were subjected to some of the torture techniques known to be used by the North Vietnamese. Our instructions were to try to tell as little as we could, and to frustrate the "enemy" in any way we could.

One of the "torture" techniques was to place you into a confined box for an extended time, interrupted only every couple hours by trips to the interrogation room. In the building there was a long wall with multiple compartments behind small cabinet doors. Each compartment had a movable partition that they could use to re-size the space to make it as small and uncomfortable for you as possible. Back then I was 6'1" as I am today, but I probably weighed all of 150 pounds. They took one look at me and slipped the partition to the smallest setting and pushed me in head-first. While I was in there, I found that I could turn around so when they opened the door a few hours later to take me to interrogation, I was facing out. That really aggravated them which, of course, was my purpose.

Another "torture" technique was to place you standing and facing the wall. You would have to extend your index fingers with the fingertips on the wall at about shoulder height. Then they would kick your feet backward until you were leaning and the only thing holding you up was your index fingers. No permanent harm done but it was very painful.

Another technique was water boarding. You had to lie on your back on a narrow bench. Several of the trainers would restrain you. They would place a wet cloth over your mouth and nostrils and then pour water onto the rag. Again, no permanent harm done but you could see how in the hands of the real enemy, this could be very effective.

Toward the end of Survival School, David and I started talking about the travel from Spokane to Vietnam. We were scheduled to fly out of Travis AFB in Sacramento only a few days after completing Survival School. It was too short of a time to go home for another visit, so we decided to see the sights in San Francisco. We got a room at the Bachelor Officer Quarters at Hamilton AFB which was just north of the Golden Gate Bridge in Marin County. We spent the two days or so visiting Ghirardelli Square, Fisherman's Wharf, and a few other sights in the area. We both remember eating boiled shrimp and sourdough bread at Fisherman's Wharf.

CO-PILOT STORIES

SETTLING IN AT CAM RANH BAY

David and I lost touch for a few days after we left Travis. We were on different flights to Clark AB in the Philippines for Jungle Survival School. We were in different training groups when there. The training consisted of some academics and a field exercise in the jungle. My brother Sam had told me stories about the pygmy Negritos who were used to "sniff" out trainees who were trying to escape and evade in the jungle. When we were there it rained torrentially so our field exercise was cancelled, and we were sent to Vietnam without that experience.

My orders were to the 536th Tactical Airlift Squadron of the 483rd Tactical Airlift Wing based in Cam Ranh Bay. David's orders were to the 459th TAS which was part of the 483rd wing but based up north in Phu Cat. When he arrived in Phu Cat, they told him not to even unpack, that the 459th squadron was being shut down and everything moved to Cam Ranh Bay.

When I arrived at the living quarters of the 536th, there was one room in the Quonset huts that had only one person in it – his roommate had rotated home recently. I was told to move into that room. On the second night, about 6 or 7 pm, there was a knock at the door. Since our room had the most recent arrival (me), we were junior to all others in seniority and so if anyone had to be three to a room it would be us. The next new arrival (David), was at the door. That's how we became roommates for the year and solidified the friendship that we had begun in Abilene and continued through Survival School in Spokane. The other guy rotated home after a couple months.

Before being occupied by the 536th TAS, our hootch had been home to an F-4 squadron. It was two long Quonset huts sitting parallel to each other with a conventionally framed building connecting them in the middle. From the air it looked like a giant "H" and it really caught your attention from the air because it was painted bright yellow. The connecting buildings housed the showers and latrine, and during the F-4 days, a squadron officer's club. The Air Force had since declared that there could be only one official officer's club on base and so that portion of the building was empty.

A few months later, David and I decided that living in a rectangular room with a flat ceiling would be better than the Quonset hut, so we took on the task of remodeling the former O-Club into three rooms and a living room area. We scrounged 2x4's to frame the walls, luan plywood to serve as paneling, doors, electrical materials, and an air conditioner or two. It seemed that no one knew where the circuit breakers were that powered that area, so we wired the 220-volt air conditioner with the power hot. In the process, the screwdriver slipped, and I received the only 220-volt shock I have ever felt.

The Cobra Cabana (yellow) at Cam Ranh Bay.

Cam Ranh Bay runways looking south to the bay in the background.

CRB commercial side ramp. Note C-5 and DC-8 Freedom Bird.

CRB military side ramp.

FIRST OPERATIONAL FLIGHT IN CARIBOUS

After a few days of orientation, going through the briefing book at squadron ops, and being issued the .38 caliber Chief Special that we carried when flying, I was scheduled to fly my first operational flight. As a Co-Pilot upgrade, I would fly with an Instructor Pilot who would teach me the ropes. I don't remember anything about the upgrade process, only that it involved several flights. We had learned most of the flying stuff back in Abilene. The challenging parts of this were learning to navigate, becoming familiar with the Tactical

Aerodrome Directory which contained all pertinent information about the destination runways, and learning how to use the radios to communicate with the air traffic control (where it existed), Airlift Control, the Army on the FM comm radio, the artillery sectors, and the personnel at the destination of our cargo load. It was no more than a few flights, but I was anxious to finish the training and get on with doing the job.

One of the more difficult skills that a new Co-Pilot had to master was managing the communication and navigation radios. We typically had one or two UHF comm radios, one or two VHF comm radios, an FM comm radio for communicating with the Army, a TACAN nav radio, a VOR nav radio and an ADF nav radio. So that was probably four comm radios that you monitored for voice transmissions and three nav radios that you had to verify the morse code identifier each time you changed frequency. On a typical sortie from a major base out to a forward firebase, you would talk in sequence to Ground, Tower, Departure, multiple artillery sectors (a new one every 5-10 minutes) and then the Army contact at the destination firebase. Everyone you are talking to or monitoring is also talking to a number of other aircraft, the activities of which might or might not be relevant to what you are doing. All the while you are monitoring Guard frequency for May Day's and helping the pilot fly the airplane. Multi-tasking indeed.

On my first flight as a fully qualified co-pilot, I

remember we took off from Cam Ranh Bay and flew past Nha Trang on our way to Ban Me Thuot. Once we got to altitude and leveled off, I remember thinking, "This is the first time in my life that I have actually performed the tasks that I have trained to perform." All through school, college, graduate school, officer training, pilot training, Caribou upgrade school, survival school, jungle survival school, and finally co-pilot upgrade training, I had been getting ready to do something but had never actually done anything. I had a deep sense of relaxation come over me and felt like I had reached a new phase of adulthood.

After the first few weeks of flying in Vietnam, I began to feel comfortable and reasonably safe. After all, Army helicopters were everywhere and if we were to have a problem with our aircraft out in the middle of nowhere, we would only have to put out a "Mayday" call on the emergency Guard frequency and some helicopter from nearby would no doubt be there by the time we hit the ground. Later in the year, as US Army units were pulled back to the larger bases and even sent back to the States, this sense of comfort diminished significantly. By the end of my year, it was reasonably unusual to see U.S. helicopters in the forward areas. I had the sense that the clock was ticking and the sooner my year was up, the better for me.

SMALL COUNTRY, BIG SKY

Although the airspace around the larger airfields was very crowded, once you got a few miles away, it was a very big sky and you were pretty much on your own. At one time in the war, Tan Son Nhut airfield in Saigon was the busiest airport in the world with more takeoffs and landings than Hartsfield in Atlanta or O'Hare in Chicago. A few miles away, a young lieutenant aircraft commander in a Caribou, not even a year out of pilot training could find himself utterly on his own. Seeing another Caribou, or for that matter any other fixed-wing aircraft up close out away from an airfield was very unusual.

Lieutenant aircraft commanders flew a lot more than many of the higher-ranking officers. The Caribou was not operated by the Air Force back in the States so the majors and colonels who were our commanders had previously flown B-52's or B-58's or maybe C-141's before coming to Vietnam. Many did not fly actual missions enough to be as proficient as we were, even with our youth and lack of experience.

The Chief of Standardization for the 483rd Wing once told a group of Lieutenants an interesting thing that stuck with me. He acknowledged that we were generally more proficient in the Caribou because we got more flying time than most of the higher-ranking officers. He described that level of proficiency as knowing how the aircraft was going to react before the event happened. His early career assignments had been in C-141's and he felt that he had that level of proficiency in the C-141. His belief was that once you achieve that 100% saturated proficiency in one aircraft, that makes you a better pilot, even after you leave that aircraft and move to another. My subsequent experience in aviation confirmed this for me.

Caribous hauled mail, ammunition, passengers, food, lumber, artillery projectiles, live animals, beer (Carling Black Label was a staple), and just about anything else you can imagine. Our passengers were US military personnel, South Vietnamese military personnel, Vietnamese civilians, Montagnard soldiers with their families, and just about anyone else who wanted a ride to where we were going. I once hauled a Toyota Cressida from Bien Hoa to Cam Rahn Bay. We were told it belonged to the mayor of Cam Rahn City.

At least 15 runways in South Vietnam were Type I or Type I Restricted for Caribous. In normal circumstances Caribous were the only fixed wing cargo aircraft that flew into these and many of the other smaller and more remote runways. Our Short Takeoff and Landing (STOL) capability was unrivaled by any other cargo aircraft operating in significant numbers in Vietnam. We shared a mission with the C-123's and C-130's but since we operated into fields that they did not and because we flew low and slow, we saw some very remote, isolated, and interesting places that most USAF crews did not see.

Lonesome Caribou, somewhere north of Bien Hoa.

Katum in the Parrot's Beak area. Fairly typical of many forward fields.

On the wall in our hootch, I kept a map of Vietnam and each time I flew a new route I would draw a line with a red pen from the origin to the destination. The line didn't represent the actual route we flew, just the takeoff and landing points. From that I compiled the following list of airfields where I personally landed during my year in country. Other Caribou aircrew might have landed at others.

RUNWAYS WHERE I ACTUALLY LANDED IN VIETNAM

No.	Name	ID	L x W	Surface	C-7 Type	Location
VA3-1	Tan Son Nhut	ASM	10,000 x 150	concrete	3	
VA3-2	Bien Hoa	BNH	10,000 x 150	concrete	3	
VA2-4	Pleiku	PLU	6,000 x 120	asphalt	3	
VA3-5	Vung Tau		4,692 x 60	asphalt	3	155 @ 40 NM from BNH
VA2-7	Nha Trang		6,166 x 144	asphalt	3	
VA2-8	Dalat/ Cam-Ly		4,400 x 60	asphalt	2	158 @ 46 NM from BMT
VA2-9	Dalat/Lien Khuong		4,845 x 125	asphalt		165 @ 57 NM from BMT
VA4-10	Quan Long		3,100 x 60	PSP	2	210 @ 63 NM from BHT
VA2-11	Phan Thiet		3,600 x 60	PSP	2	043 @ 75 NM from BNH
VA2-12	Ban Me Thuot East	BMT	5,887 x 98	asphalt	2	
VA2-13	Qui Nhon		5,081 x 98	asphalt	3	
VA4-14	Truc Giang		2,800 x 60	alum mat	2	074 at 39 NM from BHT

No.	Name	ID	L x W	Surface	C-7 Type	Location
VA2-15	Kontum		3,600 x 85	asphalt	2	360 @ 21 NM from PLU
VA4-16	Soc Trang		3,100 x 60	asphalt	2	154 @ 34 NM from BHT
VA4-17	Can Tho		3,900 x 60	asphalt	2	125 @ 4 NM from BHT
VA4-20	Vinh Long		3,000 x 80	asphalt	2	055 @ 17 NM from BHT
VA2-21	Nhon Co		4,200 x 110	asphalt	2	222 @ 52 NM from BMT
VA2-28	Phan Rang		10,000 x 102	concrete	3	
VA2-29	An Khe		7,200	concrete	2	094 @ 38 NM from PLU
VA3-30	Song Be		3,400 x 60	alum mat	2	010 @ 52 NM from BNH
VA3-31	Loc Ninh		3,100 x 60	asphalt	2	345 @ 53 NM from BNH

No.	Name	ID	L x W	Surface	C-7 Type	Location
VA4-32	Con Son		3,800 x 97	asphalt	2	146 @ 98 NM from BHT
VA3-35	Thien Ngon		2,900 x 60	PSP	2	307 @ 62 NM from BNH
VA2-39	Ban Don		2,800 x 45	laterite	2	307 @ 24 NM from BMT
VA2-42	Dak Pek		1,500 x 60	asphalt	1	351 @ 66 NM from PLU
VA1-43	Gia Vuc		3,200 x 63	sod	2	037 @ 53NM from PLU
VA3-50	Phouc Vinh		3,700 x 59	asphalt	2	356 @ 20 NM from BNH
VA4-51	Muc Hoa		2,900 x 60	laterite	2	018 @ 43 NM from BHT
VA4-52	Tra Vinh		3,600 x 100	asphalt	2	105 @ 34 NM from BHT

No.	Name	ID	L x W	Surface	C-7 Type	Location
VA4-58	Bac Lieu		2,000 x 60	PSP	2	180 @ 47 NM from BHT
VA3-61	Sanford AAF		3,200 x 60	asphalt	2	
VA2-89	Buon Tsuke		1,800 x 80	sod	2	203@23 NM from BMT
VA2-97	Duc Xuyen		1,300 x 60	PSP	1	178 @ 28 NM from BMT
VA2-107	Bu Prang New		2,400 x 60	asphalt	2	027 @ 84 NM from BNH
VA2-113	Tuy Hoa		9,500 x 150	concrete	2	
VA3-121	Bu Dop		2,900 x 60	laterite	2	360 @ 63 NM from BNH
VA3-129	Dau Tieng		2,500 x 60	laterite/ asp		305@ 32 NM from BNH
VA3-130	Djamap		3,700 x 60	laterite	2	017 @ 70 NM from BNH

No.	Name	ID	L x W	Surface	C-7 Type	Location
VA3-131	Gia Ray		1,500 x 60	laterite/ sod		090 @ 36 NM from BNH
VA3-132	Ham Tan		3,400 x 80	laterite	2	107 @ 57 NM from BNH
VA3-135	Lai Khe		3,500 x 60	alum mat		319 @ 17 NM from BNH
VA3-147	Quan Loi		3,900 x 90	laterite		348 @ 42 NM from BNH
VA3-153	Phu Loi		2,800 x 85	asphalt	2	282 @ 7 NM from BNH
VA3-159	Vo Dat		3,700 x 156	laterite/ penep-rime	2	076 @ 41 NM from BNH
VA4-167	Rach Gia (Kien Giang)		1,300 x 100	asphalt	2	256 @ 35 NM from BHT
VA4-173	That Son		2,000 x 53	PSP	2	303 @ 49 NM from BHT
VA 4-187	Binh Thuy	BHT	6,000 x 100	asphalt	3	

No.	Name	ID	L x W	Surface	C-7 Type	Location
VA2-192	Cam Ranh Bay		10,000 x 150	concrete	3	
VA4-193	Chau Duc		1,600 x 98	sand/ clay	2	317 @ 49 NM from BHT
VA2-202	Gia Nghia		2,100 x 65	laterite	2	213 @ 46 NM from BMT
VA2-213	Phu Cat	PHJ	10,000 x 150	concrete	3	
VA4-225	Vi Thanh		2,300 x 60	psp	2	216 @ 22 NM from BHT
VA3-226	Xuan Loc		3,500 x 80	asphalt	2	097 @ 26 NM from BNH
VA2-232	English		3,600 x 60	asphalt	2	358 @ 32 NM from PHJ
VA2-241	Polei Kleng		3,500 x 50	psp	2	333 @ 27 NM from PLU
VA3-248	Tanh Linh		2,000 x 60	psp	2	083 @ 52 NM from BNH

No.	Name	ID	L x W	Surface	C-7 Type	Location
VA3-256	Tay Ninh West		3,800 x 80	asphalt	2	296 @ 49 NM from BNH
VA2-260	Bao Loc		3,500 x 59	psp	2	196 @ 69 NM from BMT
VA4-264	An Thoi		3,600 x 60	psp	2	260 @ 100 from BHT
VA 2-265	Tieu Atar		1,500 x 60	laterite	1	330 @ 39 NM from BMT
VA3-269	Duc Phong		3,000 x 60	laterite/ penep-rime	2	027 @ 56 NM from BNH
VA3-278	Di An		2,800 x 90	laterite/ penep-rime	2	220 @ 6 NM from BNH
VA2-283	Dak Seang		1,400 x 45	clay	1	338 @ 53 NM from PLU
VA3-284	Luscombe		2,900 x 80	asphalt	2	135 @ 35 NM from BNH
VA3-287	Katum		3,000 x 80	laterite	2	319 @ 55 NM from BNH

No.	Name	ID	L x W	Surface	C-7 Type	Location
VA2-293	Duc Lap #2		3,300 x 60	asphalt	2	243 @ 30 NM from BMT

The first number in the Aerodrome Number indicates the Corps, i.e., I, II, III, or IV Corps

Smaller fields are located with radial and distance from larger fields.

Larger fields have TACAN identifiers.

PSP = pierced steel plank

Laterite = red clayey soil/weathered rock that is hard when dry

Laterite/peneprime = laterite treated with oil

Information from Tactical Aerodrome Directory South Vietnam dated 17 AUG 1972

AF ACADEMY A/C AND TACTICAL APPROACH TRAINING

Shortly after I arrived in-country, the Caribou squadron at Phu Cat was brought to Cam Rahn Bay so for most of my year, all four squadrons operated out of CRB. We had two regular temporary duty locations in Bien Hoa and Can Tho.

For Bien Hoa, we would fly out of CRB (with clothes and personal gear for a week) in the morning, fly a full day of missions, and recover in Bien Hoa. We would then fly two days out of Bien Hoa, have a day off, and then two more days. On the seventh day, we would launch out of Bien Hoa (with our personal gear on board), fly a full day of missions, and recover in CRB. We called it the "seven-day stage." A number of Caribou crews would be in Bien Hoa at the same time, so we had quarters designated for our use.

For Can Tho, we had a similar routine, but it was only three days and no day off. We lived in a civilian hotel downtown while there. The best meal I had in Vietnam was chateaubriand with avocado salad and strawberry shortcake at a small French restaurant down the street from our hotel in Can Tho.

Although these TDY's took us away from our permanent quarters and our personal stuff, they were a nice break in the routine. Cam Rahn City was closed to us. On our day off at Bien Hoa we could take a taxi to Saigon to go shopping, eat at the Continental Hotel, and use the MARS phone at the USO to call home. In Can Tho we usually had a few hours in the evening to walk the streets, have a nice meal, and have a drink or two.

There was a fair amount of fluidity in the crew assignments. A crew of aircraft commander, co-pilot, and flight engineer might fly together for a few days or a few weeks or even a month or two, but it wasn't unusual for crew assignments to be shuffled on a regular basis. Not long after coming in country and checking out as a co-pilot, I was assigned to fly with one of the seasoned lieutenant aircraft commanders. Right away we were sent to Bien Hoa on the seven-day stage.

As I remember it, this A/C was a graduate of the Air Force Academy and his father was an Air Force 3-star general. He was well-schooled in Air Force life, very intelligent, and very cocky. He was intolerant of silliness and stupidity, took grief off of no one, and occasionally walked a very fine line just short of insubordination.

We had been taught to perform tactical approaches at Dyess in Abilene, but this guy had his own refinements to that procedure and at Bien Hoa he went right to work teaching them to me. At the core of his technique was the principle of trading off potential and kinetic energy. Since my education was in engineering, he found me to be a receptive pupil.

His technique was to arrive over the destination field at about 3,000 feet above ground level, with the aircraft at minimum airspeed. This brought the sum of your potential and kinetic energy to a minimum while keeping you out of the range of small-arms fire. You would then enter a steep spiraling turn with engines in idle, descending to the landing pattern elevation and arriving on the downwind leg at about 120 knots. With throttles in idle and propellers set for cruise, the propellers acted as huge air brakes, preventing the airspeed from getting too high in the steep spiraling descent. This maneuver dissipated energy and converted the remaining energy to kinetic energy in the form of airspeed. At this point, you had to increase throttle briefly just long enough to set the propellers for a possible go-around and then return them to idle. Then flaps, landing gear, and final turn with minimum use of power. Ideally, you could make the final approach and landing without using *any* throttle.

The guiding principle behind this technique was avoiding noise. This A/C's idea was that "Charlie" should not hear us until we added power for takeoff. That would deprive him of the opportunity to mortar us while on the ground and minimize the time he had to set up to take shots at us on the way out.

I am sure that other Caribou pilots had their own techniques but this one made sense to me and it gave me a real feel for what was going on with the airplane during this critical phase of flight.

I used this technique for the balance of my tour. Once up north near Quin Nhon (maybe at An Khe) when I demonstrated this to a new co-pilot, the controller on the ground asked how many of our passengers got sick. He was right that the sudden rapid descent did have the potential to put your stomach in your throat if you didn't know it was coming.

ALWAYS FOLLOW THE RULES

That same week we were in the Caribou Operations area one morning preparing to start our day. The Major that was in charge in Bien Hoa said something to the Aircraft Commander (the AF Academy graduate) to the effect that he better not hear of us doing "xxxxxx" because that would be a violation of the rules. The A/C turned to him and asked, "What did you say?" The Major repeated his admonition to not break the rules. The A/C replied, "Do you mean that we are supposed to comply with all the rules, no matter what?" The Major replied in the affirmative.

We flew the entire day without delivering one single load. We would load the aircraft at Bien Hoa, fly to the destination, identify some reason why we were not allowed to land (weather, security, airfield condition, etc.) and then return to Bien Hoa. We would unload at Bien Hoa, pick up the load for the next sortie, and repeat the whole process. This was obviously ridiculous and a bit extreme, but it made the A/C's point.

We operated under a whole list of ridiculous rules. After a few months, I came to realize that the purpose of the rules was to shift the responsibility for accomplishing the mission under difficult circumstances to the young, mostly lieutenant pilots. In case something unfortunate happened, the higher-ranking commanders who made the rules were protected from blame. It could always be said that the young, inexperienced pilot should have known better than do "xxxxxx," because, after all, it was against the rules.

In the vast majority of circumstances, we managed to get the job done competently and safely, despite the rules.

WOMAN AT LAI KHE

Early in my tour we were flying the Bien Hoa seven-day stage. One day we landed several times in Lai Khe, a small airfield only about 20 or 30 miles northwest of Bien Hoa. The ramp area we used there for offloading was more or less deserted. As I remember it, there was a small building at the edge of the ramp. The building had been damaged by ordnance of some sort and there was little else around. Although the area was relatively secure, we usually did not shut down engines for offloading so at least one and usually both pilots stayed in their seats the entire time.

Each time we would arrive, a Vietnamese woman would come out and stand in front of the aircraft. She looked old but might have just had a hard life. She would lift her dress up to reveal her naked body underneath.

It was not at all clear what she was trying to communicate but the story we were told was that she had syphilis. It is quite possible that she was in the tertiary stage of syphilis which can cause insanity. Whether she was a victim of the war or a victim of her own poor judgement, we couldn't know, but she was someone's daughter and possibly sister and mother. There were many sad circumstances in Vietnam and hers was certainly one of them.

SANTA BOU AT DJAMAP

Each Christmas, the maintenance crews would paint a Santa Claus face on the nose of several Caribous. It actually worked pretty well because the nose of the fuselage was shaped more or less like a head and the radome made a believable nose. Each of these aircraft would be assigned a few Air Force personnel and a couple of Red Cross Donut Dollies. With some donuts, coffee, eggnog, and other Christmas goodies, they would fly to some of the forward firebases and spend a few minutes bringing Christmas cheer to the guys in the field.

Our normal call sign was "Iris" followed by our mission number, maybe something like "Iris 416." For Christmas day, these painted Caribous would use the call sign "Santa Bou 420" or something like that.

On Christmas morning, 1970, our first sortie was to Djamap, a firebase north of Bien Hoa, just a short distance from the Cambodian border. We left Bien Hoa early and when we reached Djamap it was overcast, foggy, and really not full light yet. Fifteen or twenty minutes out, we tuned the Comm radio to the Djamap frequency and just listened to see what else was going on.

"Santa Bou 420 in-bound Djamap for landing."

No response from Army guy at Djamap.

Then again after a few minutes, "Santa Bou 420 in-bound Djamap for landing."

A long pause and then a very tentative "Roger."

This Army guy was obviously spooked by the Santa Bou call sign and seemed unsure whether his mind was playing tricks on him.

We waited a few minutes and then called "Iris 416 in-bound Djamap for landing."

An immediate response, "Roger, Iris 416."

And then, "Iris, did you hear someone else on this frequency a few minutes ago?"

After a good laugh, we explained it to him. When we landed, Santa Bou was on the ramp spreading their Christmas cheer. We unloaded, visited with them for a few minutes, and took off. As we turned back toward Bien Hoa, Santa Bou was still there.

Santa Bou at Djamap.

Santa Bou still there as we leave Djamap.

HAM TAN FOR CHRISTMAS

On Christmas Day, 1970, we were flying the seven-day stage out of Bien Hoa. The weather was mostly overcast, and we drew an afternoon mission to Ham Tan, a small army installation due east of Bien Hoa, near a small river and only a few miles inland from the coast. Our load was a conglomeration of mail, holiday foods (fruit, turkey, etc) and some miscellaneous cargo. There was a high cloud deck above us and a low overcast below from about 3,000 feet down to within a few hundred feet of the ground. All in all, a really gloomy day

When we arrived over Ham Tan, we contacted the Army on the ground and advised them that we would not be able to land in those conditions. There were no navaids and certainly no approach aids into those small fields. The Army guy answered, "Oh Man—we haven't had mail for five days and it's Christmas." We told him we would see what we could do.

We climbed a few hundred feet, maybe to about 4,000 or 4,500 feet in the clear between the two decks of clouds. At that altitude, we had a good lock on the Bien Hoa TACAN and could accurately position ourselves. We headed east until we were certain we were well out over the ocean. The Aircraft Commander flew a shallow spiraling descent on instruments and as the Co-pilot I cross-checked and watched out the windscreen. The Flight Engineer was standing between the pilots with instructions to concentrate on picking up the first sign of the water below us. We would spiral down until we reached 400 or 500 feet and if we didn't break out of the bottom of the clouds, we would level off and climb due east away from land until we broke back out between the cloud layers.

The bottom of the lower deck was ragged and so we started picking up the ocean surface well before our go-around altitude and were able to easily level our wings and turn back to the west toward the coast. On reaching the coast, we flew north until we found the mouth of the river. Staying as high as we could and maintain visibility (maybe 500 feet), we followed the river upstream to Ham Tan. Inland, the ceiling was a bit higher and we were able to pull up to about 800 feet to fly a downwind and land.

On the ground was a deuce-and-a-half full of Army guys cheering and waving at us. We had a lot of help unloading. I had always understood the purpose of the Caribou operations in Vietnam as supporting the ground troops. After that day, I had a new appreciation for the role we could play.

LAUGH BOX AT BIEN HOA

Each morning at Bien Hoa, we would pre-flight our aircraft about dawn and taxi to the cargo ramp to pick up our first load for the day. A half dozen or so Caribous would be doing the same thing and multiple other aircraft would be starting their daily routine as well. Bien Hoa ground control monitored the movement of all of these aircraft, and you would request permission to taxi from the Caribou ramp to the cargo ramp and then to the runway for takeoff. For that reason, for a period of time each morning, all of the Caribou aircraft and many others would be tuned to the Bien Hoa Ground Control frequency. By 1970-71, many air traffic control functions, particularly the less critical functions such as ground control, were being turned over to South Vietnamese controllers. The operational control language was English but many of the South Vietnamese controllers spoke English with a heavy accent.

One morning someone keyed their microphone and activated a "laugh box" on ground control frequency. These were novelty items available from many sources in country. As I remember them, they were a small plastic box with batteries and a speaker. They were usually sold with a small cloth sack with a drawstring. When activated, they played audio of continuous uproarious laughter.

That morning, each time Ground Control would transmit instructions to an aircraft, an unidentified aircraft would key the mike and transmit a few seconds of laughter. As I remember it, it went something like this.

Ground Control: "Aircraft XYZ, taxi and hold short of runway 27 Left"

Mystery Aircraft: "Ha Ha Ha Ha Ha Ha Ha Ha Ha Ha"

For the first few times, the Ground Controller tried to ignore the laughing but at some point, he could no longer ignore it.

Ground Control: "Aircraft transmitting on Bien Hoa Ground, say call sign."

Mystery Aircraft: "Ha Ha Ha Ha Ha Ha Ha Ha Ha Ha"

Ground Control (more emphatically): "Aircraft transmitting on Bien Hoa Ground, say call sign."

Mystery Aircraft: "Ha Ha Ha Ha Ha Ha Ha Ha Ha Ha"

Ground Control (even more emphatically): "Aircraft transmitting on Bien Hoa Ground, say call sign!"

Mystery Aircraft: "Ha Ha Ha Ha Ha Ha Ha Ha Ha Ha"

Ground Control (now clearly losing his patience): "Aircraft transmitting on Bien Hoa Ground, SAY CALL SIGN!!"

Mystery Aircraft: "Ha Ha Ha Ha Ha Ha Ha Ha Ha Ha"

And then from another unidentified aircraft, very dryly: "He might be crazy, but he's not stupid."

It clearly was disrespectful of the Ground Controller, but it was some much-appreciated comic relief at the start of our day.

ARVN FAMILY AT LOC NINH HEADED TO SAIGON

One day we were on a routine flight from Saigon to Loc Ninh and back. At Loc Ninh we took on passengers, among them was a young ARVN lieutenant and what were evidently his wife and child. It struck me that his circumstance and mine were in some ways parallel. I also had a wife and young child. My good fortune was that if I survived my year in Vietnam I would go home to a land of law and order and security and opportunity. His family's future was much more uncertain.

I have wondered many times since how the subsequent betrayal of our Vietnamese allies by the United States Congress affected that young man and his family. I can hope that they were some of the thousands that escaped South Vietnam to live in the US or elsewhere. Unfortunately, the odds are against that outcome.

STAR SPANGLED BANNER AT CRB

At Cam Ranh Bay, the Post Office was on one side of a square bordered by streets. At the center of the square was the flag pole. On a day that I was not flying I had walked down to Squadron Operations for some reason and was walking up to the Post Office on my way back to the hootch. I was dressed in my jungle fatigues.

As I approached the square, the color guard had arrived to lower the flag for the night. As the Star Spangled Banner began over the loudspeaker, I stopped and saluted while the flag was lowered. I remember thinking, "Oh, this is what it is like to be in a foreign land serving your country." I think I had been so busy trying to do all the things that were required of us that until that moment, I had not experienced that moment of self-consciousness.

LOBSTER FROM PHAN THIET

Phan Thiet was a coastal town south of Cam Ranh Bay and more or less due east of the Saigon/Bien Hoa area. Pretty much every day a Caribou mission would fly to Bien Hoa and back, perhaps completing a few sorties in the Bien Hoa area before returning. It was not unusual that on the way down and back, there would be stops at Phan Thiet. The preferred route would be "feet wet" (over the ocean, just off the coast) down the coast to Phan Thiet and then due east to Bien Hoa. That avoided the early morning fog and low clouds over the mountainous terrain.

By the time I arrived in country it had become a squadron tradition to have a lobster boil every few months. The aircraft headed for Phan Thiet would be loaded with several empty 55-gallon drums that would be dropped off in Phan Thiet. On the return that afternoon, they would pick up those drums filled with small lobster that had been caught that day.

Back at the squadron hootch, in anticipation of the return flight, a fire would be built under a drum of water and the "fixin's" would be prepared, baked potatoes, French bread, and perhaps a salad. I think the cost was something like 10 cents per lobster with all you can eat fixin's for free. I am sure there was iced tea, but I think the principal beverage was beer.

AIRCRAFT COMMANDER STORIES

LANDING AT DUC XUYEN

We were flying a normal day's mission out of Cam Rahn Bay. We had flown to Ban Me Thout where we picked up a load for Duc Xuyen. None of the crew had ever been there.

It was mid-afternoon and there were puffy cumulus clouds and small rain showers scattered all over. As we approached Duc Xuyen, a shower moved across the runway, so we just maintained our altitude and loitered for a while until the shower had moved through.

The runway at Duc Xuyen was Pierced Steel Plank, PSP. These were channel shaped panels of steel with holes about three inches in diameter in the upper surface. They interlocked so that when placed side by side, open side down, locked together, and filled with dirt they formed a reasonably good runway surface in forward areas. Duc Xuyen didn't see a lot of traffic and so grass had grown in all the piercings.

Wet grass over a hard surface can be slippery and that day it was.

Normal procedure at touchdown was to tap the brakes to ensure that they were functional and place the propellers in reverse. We used the reverse thrust to slow down and only used the brakes at the end of the roll-out. On really short fields you might pump the brakes along with reverse thrust.

After the shower had moved through, we made our approach and landed. When I tapped the brakes, the left main locked up and start sliding. The left side got ahead of the right and we were sliding down the runway at a scary angle, although it probably seemed like more than it was. I added power to the left engine and the reverse thrust brought the nose back around.

The Caribou was a tough bird but side-loading the landing gear is asking for trouble. Sometimes there was only a hair's breadth between "ho-hum" and disaster.

Young Boys at Duc Xuyen.

Runway at Duc Xuyen.

TRIP TO RACH GIA WITH BRASS ON BOARD

We were flying the seven-day stage out of Bien Hoa on a totally ordinary day. We flew a sortie or two in the morning and arrived back at Bien Hoa before lunch. Our next sortie would be to Rach Gia, about 150 miles southwest of Bien Hoa. I don't remember the cargo. I had never been to Rach Gia but that was not unusual -- we frequently were sent to airfields that were unfamiliar. That was one of the interesting and challenging things about being a Caribou pilot.

While we were waiting for the load to arrive, the OIC (Officer In Charge) of the Bien Hoa operation, a major, arrived and announced that he would be riding with us. Not a formal check ride or anything, just along for a friendly ride. He was a nice guy and I was confident and not at all intimidated.

We had started the day with full fuel tanks and had flown only a couple of short sorties. We normally operated with less than full tanks because full fuel tanks limited our cargo capacity and performance significantly. I didn't think we would need fuel until we returned from Rach Gia. The Major asked if we were going to take on fuel and I said I thought we would be ok.

We left Bien Hoa, delivered our cargo to Rach Gia, and took off to return to Bien Hoa. As soon as we reached cruising altitude and things settled down, I realized that we were bucking a significant headwind and our ground speed was miserable. It didn't take long for it to be painfully obvious that we should have taken on fuel and that I had made a serious error with the Major on board.

I reasoned that the worst thing I could do would be to compound my error by refusing to own up to it. We would have to stop somewhere to get fuel. I called Airlift Control in Saigon and asked them if they had a load in Binh Thuy that needed to go to Bien Hoa. Standby. A few minutes later, sure enough, they did have such a load. We diverted slightly off our course to Binh Thuy, took on fuel and picked up that load. Back in Bien Hoa, the major didn't say much when he got off the plane. I think he was amused by it all.

A few months later, I was offered the opportunity to upgrade to Instructor Pilot. This was quite an honor because, at any one time there were only a few Lieutenant Instructor Pilots in the Wing. I don't know if this incident had anything to do with being selected for IP upgrade, but I was lucky to have flown with a commander who understood that a mistake recognized quickly and properly corrected is a good learning experience.

SEA SURVIVAL SCHOOL AT KADENA

Around the middle of January 1971, I was sent to Okinawa to attend Sea Survival School at Kadena AB. As I remember it, the course lasted one week. It began with the academic portion and concluded with a simulated bailout and rescue.

Kadena in January is not exactly a South Pacific paradise island. The day we went out for the simulated bailout, the air temperature was about 40 degrees and the wind was blowing about 30 knots. I have no idea what the water temperature was but let's just say it was COLD.

The Air Force had converted a World War II landing craft to launch the students on a parasail. There was a large screen on one end of the craft. They would throw the parasail up against it and the wind would hold it there while they attached it to your parachute harness. Clothing was a cotton flight suit, tennis shoes, and a helmet like the ones we flew in. Once hooked up, a speedboat would come alongside and up you would go. At several hundred feet, they would cut you loose to simulate a parachute landing in the water.

We were taught to determine when we were about to hit the water and disconnect our parachute from the harness before hitting the water so that we wouldn't get tangled in the parachute. Then inflate the one-man raft, turn on the locater beacon, put out the shark repellent, try to stay as warm as possible and eat the C rations that were in the survival kit. After about 30 minutes they came by to pick you up.

I have never been so cold in my life. After being picked up, they returned you to the landing craft so you could wait until everyone in the group completed the simulation. There was a big pot of hot bouillon or something similar, but I really didn't warm up until the next day.

The cassette tape was the new music format at that time, so I took the opportunity to shop at Kadena for stereo gear, including a cassette tape deck. The newest thing in reel-to-reel decks was Akai's glass and crystal ferrite record & playback heads. Akai also had a very nice looking new reel-to-reel deck that was designed to sit on a shelf with other stereo gear. Unfortunately, the new glass heads were not available in this new furniture-grade tape deck. I decided to buy it anyway and David told me to buy him whatever I got for myself, so I bought two.

Within a week or so after returning to Cam Ranh Bay, Akai announced a new version of their new tape deck and it had the glass heads. I was sick. I was depressed. I was heartbroken. What lousy timing.

David immediately decided to put his brand-new tape deck on the bulletin board, someone bought it, and he put the money aside to buy the new model with the glass heads as soon as he could. I wondered to myself why I did not react that way. Eventually I did the same and was very happy with the result and grateful to David for teaching me a lesson. I think my upbringing as the son of a widow with very limited money had taught me to find a way to "make do" with whatever I had. I had seen a glimpse of another way to live.

Me at Sea Survival School in Kadena, Okinawa.

CON SON ISLAND

One day we were operating out of Saigon and were assigned a load for Con Son island, about 120 miles due south of Saigon. About half of that distance was over open ocean. At the time, Con Son island had been in the news as the location of the prison where "political prisoners" were being held in "tiger cages."

The weather was excellent and except for the long distance over the ocean, the flight out and back was totally uneventful. Caribous frequently flew for long distances over the ocean but usually within a few miles of the coast, say from CRB along the coast north to Tuy Hoa or Quin Nhon or south to Phan Thiet. Flying straight out into the ocean was not our usual thing.

As I remember it, the runway stretched from beach to beach at a narrow spot of the island. It was very quiet there, so we shut down and found a Navy chow hall to have lunch. We always thought the Navy had better food than the Air Force and that day we were not disappointed.

AN THOI ISLAND

We were flying the three-day stage out of Can Tho and we drew a mission to An Thoi. An Thoi was on an island due west of Can Tho but far enough out into the Gulf of Thailand that you would think it was part of Cambodia. It was about 90% surrounded by Cambodian waters and was actually southwest of Phnom Penh.

The weather was nasty, with overcast skies and rain. We left Can Tho and headed to Rach Gia, located on the west coast of Vietnam. Over the delta, we had no problem maintaining a safe altitude.

Over Rach Gia, we turned due west toward An Thoi. Over the ocean the rain got harder and the ceiling lowered dramatically. We descended to about 300 feet where we could maintain some visibility. Unfortunately, at that altitude we lost the Binh Thuy TACAN and so navigation got a bit iffy.

From Rach Gia to An Thoi was about 60 miles and about 30 miles out, right on line with our course was a small island. We adjusted the weather radar to sweep level with our altitude and easily picked up the island as a dark shadow on the radar screen. We flew straight at that shadow, using the radar to navigate. As we eased around the side, the highest terrain on the island was above our altitude.

On the other side, we resumed our previous heading and used the radar again to paint the island where An Thoi was located. The landing was uneventful. We used the same procedure on the return trip.

At An Thoi, I photographed a Vietnamese firefighter crew and their American issue fire truck. It looked huge next to the firemen who were small in stature. They obviously had been issued USAF fire suits that were much too big for them. They could have been little boys playing dress-up but they were all smiles for the camera.

Firemen at An Thoi Island.

ARTILLERY NEAR MISS AT LOC NINH

When you arrived in country, part of the orientation was to spend some time at squadron operations reviewing a package of information about Caribou operations in Vietnam. One of the documents was the photograph of a Caribou on short final that had just had its tail section blown off by friendly artillery. The fuselage was pointed straight at the ground, about 50 feet up. You had to wonder what the pilot was thinking as he faced certain death. There is a photo of this incident posted on the C-7A Caribou Association web site.

One day when we were flying the Bien Hoa stage, we were headed to Loc Ninh, about 50 miles north of Bien Hoa. As we flew along, the co-pilot checked with various artillery sectors and firebases to find out whether they were firing and if so, in which direction and to what altitude. We routinely had to adjust our course to avoid active artillery. As we approached Loc Ninh from the south for landing, we checked with the artillery controller at Loc Ninh and he said they were firing to the west. The firebase was just west of the runway, so we set up our downwind on the east side and were going to make a left-hand final turn to the south to land.

Loc Ninh had been a rubber plantation and just west of the runway, on the northern half of the runway was a mansion with swimming pool and tennis court. The place was abandoned and in disrepair, but you could tell that in some former time, life had been very good for someone in Loc Ninh. As I was looking out the open window at this mansion, the aircraft shuddered pretty violently, a feeling somewhere between a vibration and the jolt you would get from heavy turbulence.

I looked back under our left wing and saw to my horror the long barrel of an artillery piece pointed right across our tail, smoke rising from the business end. By the time we realized what had happened, it was over, and it was obvious we were ok.

Loc Ninh mansion.

Loc Ninh runway & firebase.

When we taxied into the ramp area, a US Army Lieutenant came rushing up. He apologized for the ARVN artillery mistake and tried to explain. I was too relieved to be alive to really be mad. I kept thinking about that photograph at Squadron Ops and what a fine line is often drawn in wartime between disaster and a good survival story.

THE BIGGEST RED ONE

The Army had land-clearing crews whose job was to clear large swaths through the trees to deny the enemy the cover of uninterrupted jungle. They used large bulldozers with "Rome Plow" blades specifically designed for taking down large trees. It seemed that these were particularly common in the area north from Bien Hoa to the border.

One day we had flown a sortie to one of the firebases along the border in that area, when we encountered these large cleared strips in the jungle except that these had been laid out in the shape of the shoulder patch of the Big Red One. I had to take a picture. I really don't know the dimensions of this thing, but I would bet it was several miles tall.

P.S. The Vietnam Land Clearing Association has a museum in Cedartown, Georgia. Rome Plow blades were manufactured in Cedartown.

The Biggest Red One.

KOREAN PILOT AT SAIGON

For a while during the war, Tan Son Nhut airport in Saigon was the busiest airport in the world. Anyone who flew in and out of there knew it was busy. In addition to the volume of aircraft, there was a problem with the variety of aircraft. Mixing O-1 Bird Dogs, Caribous, F-4's, C-130's, all sorts of helicopters, and commercial airliners of all ages, types, and nationalities seems in retrospect to have been a recipe for disaster. It is amazing that there were not more accidents.

The runway we used was 10,000 feet long by 150

feet wide with intersecting taxiways only at the approach and departure ends, a virtual playground for the Caribou. The taxiway out to the approach end of the runway actually intersected the runway a few hundred feet from the end. Rather than land normally, using all of the runway and having to taxi back such a long distance, we would frequently (traffic permitting) land short (never on the overrun of course, wink, wink) and then taxi to our ramp via the taxiway that departures used to taxi from the ramp out to the runway.

We were on short final one day, contemplating the possibility of doing just that when I noticed a South Korean single engine taildragger (maybe a DeHavilland Otter) approaching number one for takeoff. After we were cleared to land and just as I concluded that we would have to use the whole runway to avoid a head-to-head with him on the taxiway, he pulled out and took the runway for takeoff. We were less than 1/4 mile final.

Having plenty of runway to work with, we just added power, flew over him, and landed in front of him. Taxiing in, we were laughing about whether he was on a check ride and whether he would pass or fail after taking the runway without clearance from tower. We pulled into the Caribou ramp area, shut down the engines, and began unloading the aircraft.

Next thing we know, here came two military police jeeps, full of MPs and one poor disheveled Korean guy in a flight suit. They pulled up to our aircraft and the guy in charge came over to me, identified himself, and asked if I wanted to "press charges" against the South Korean pilot. I declined, fully confident that the stern discipline he had already received at the hand of the ROK military would make him remember that day for years to come.

KIA AT PHOUC VINH

Caribous were occasionally called on to transport KIA's. These were always sobering experiences. There were a few special procedures, but mainly the principal we followed was that KIA's got priority over anything else coming back from the forward areas.

We were flying the seven-day stage out of Bien Hoa and were basically shuttling back and forth between Bien Hoa and Phouc Vinh, about 30 miles or so north. It was not unusual for three or four Caribous to spend the day just flying back and forth between two locations such as these, usually a major rear area base and a smaller forward base.

While unloading on one of our first sorties that day, the Army at Phouc Vinh let us know that there would be a KIA for us on the next run. We left for Bien Hoa and in an hour or so were back. We inquired about the KIA and they said, "He's not ready yet, call us next time."

We told them we would be back and forth all day and to just let us know. Each time we landed at Phouc Vinh, we would ask, and they would respond, "He's not ready yet."

This happened maybe four or five times until mid-afternoon when they said, "We're bringing him out." A few minutes later a guy in a jeep drives up and gets out with a small box, about four inches high, four inches wide, and six inches long.

The story was that this fellow had been killed inside a tank or armored personnel carrier and that in the process of dismantling the vehicle, they had found a fragment of his skull. I think most of the body must have been sent on earlier.

We flew him back to Bien Hoa with all of the solemnity that we would have afforded a complete body in a body bag. He deserved that and much more.

OIL LEAK AT TAN SON NHUT

At Tan Son Nhut airfield in Saigon, the Caribou Wing maintained a small ramp area for four or five aircraft. Each morning an aircraft mechanic from Bien Hoa would fly over on one of the first flights and spend the day taking care of minor problems that arose on Caribou aircraft transiting that ramp. He had a small stock of commonly needed parts and his tools.

The Caribou had radial piston engines, Pratt &

Whitney R-2000's. There were two rows of seven pistons each. Pushrods from the hub of the engine extended out around the cylinders to actuate rocker arms on each individual cylinder head. Rocker arms rotated around a pivot pin that was held in place with a nut and the nut was safety wired to keep it from coming loose. As with most radial engines, oil leaks were normal.

I pulled into the ramp one day with passengers, shut down and unloaded. We were scheduled to leave about an hour later with more passengers. The mechanic on duty took a look around the airplane as he always did and spotted oil dripping from the cowling under the left engine. He pointed it out to me and said, "Lieutenant would you like me to open that cowling and check out that leak?" I took one look, thought to myself, "They all leak" and then said to him, "Sure, suit yourself."

A few minutes later, I got cold chills when he showed me that the safety wire had broken, the nut had backed off, and the rocker arm pivot pin had backed almost halfway out. Just a little more and we would have lost that engine. I will always believe that the engine would have failed during takeoff on the next sortie. At the very least I would have had a difficult emergency on my hands. More likely, the consequences would have been much worse.

I'm tempted to say that I was lucky that day, but it really had nothing to do with luck. My good fortune was the direct result of a conscientious aircraft mechanic who somehow sensed that this oil leak was not like all the others. I will owe him forever.

Tan Son Nhut Runways.

LANDING SHORT AT SANFORD

Sanford was a small Army field only about 10 miles from Bien Hoa and Saigon. On a trip from Sanford to either of these larger bases, you literally would not clear one traffic pattern before you had to enter the other. One day our mission called for us to land at Sanford and then go on to Tan Son Nhut in Saigon.

Although the Caribou was designed and built to optimize its STOL (short takeoff and landing) capabilities, many of the runways in Vietnam did not require this capability. At 3,400 feet long, Sanford was one of those. However, we were encouraged to practice good STOL techniques when we made routine landings and specifically to try to put the aircraft down on the first few feet of the runway. Basically, you would be training on every landing to make that occasional one where the runway was only 960 feet long and 40 feet wide.

As we approached Sanford, it was evident that they had just put a new asphalt topping on the runway. I did not think this through thoroughly enough to realize that there would be a 2-inch ledge at the end of the new asphalt, instead of a smooth transition to the overrun that you might expect. When I landed, I was a little short and the main gear must have touched down right on that ledge because it was a hard landing—the kind where you look out the window and are surprised that everything is still there. The Caribou was a tough airplane, but this was a hard landing. I was embarrassed, but on inspection nothing obvious was damaged, so we picked up our load and took off for Tan Son Nhut.

After a routine landing at Saigon, we were taxiing toward the Caribou ramp. We had only two right turns and a short taxi to go when I heard a metallic "plinka - plinka" sound from the left side. By the time I looked out, the outboard tire on the left landing gear was flat and flopping around on what was left of the wheel. A large chunk of the wheel had held on as long as it could and had finally broken completely out.

Lightly loaded, with the other main on that side looking normal, and with a short distance to go, we continued our taxi to the ramp.

Fortunately, the ramp mechanic had a spare wheel/tire, so he put it on. After a much closer inspection of the other three mains, we continued our mission. My lucky day.

R & R IN HAWAII

Sheri and I planned to meet in Hawaii for my R & R. I flew to Hickam on a Freedom Bird and she flew from Alexandria to Dallas/Ft Worth where she boarded a Boeing 747 for the flight from there on to Hickam. The 747's were fairly new, and although she had flown to Japan and back for summer missions, the trip was quite an experience for her.

Sheri and I discussed whether she would bring Alan. He was about a year old at the time. Left to me, we might have included him, but it was clear to her that he should be left at home with my Mom. She was right.

Our hotel was right across the street from Waikiki Beach. We rented a car and drove around the island to see some of the normal "tourist" sights. We had dinner at the club where Don Ho did his show. We went to a zoo. We visited a cultural center where they demonstrated the hula dance. We visited a Marineland type park. Mostly we just enjoyed being with each other and the respite from the different circumstance we each were dealing with. She was trying to teach science to 9th graders and I was flying airplanes in a war zone.

As when I had left home six months before, she did not ask me to promise that I would make it home. By this time, I was much more comfortable with the risks we were taking in Vietnam so I might have been tempted to make that promise. I think she was realistic enough to know that nothing is 100% certain in a war zone.

My first year in graduate school in Texas I would drive every other weekend back to Louisiana Tech in Ruston to see her. During that year, it had become harder and harder to leave on Sunday to go back. I had begun telling her, "I can't come back until I leave." I tried that same line when we had to leave Hawaii, but it didn't seem to make it any easier.

Sheri at R & R in Hawaii.

Me at R & R in Hawaii.

MAINTENANCE ALERT

Each day one crew would be designated for Maintenance Alert at Cam Ranh Bay. You would brief and pre-flight the airplane and then just sit around all day. If a crew had a problem with their airplane, the Maintenance Alert crew would fly to the location with mechanics, parts, and tools. After unloading, the operational crew would take the good airplane and continue their mission. The Alert crew would wait around with the mechanics while the broken bird was fixed and then fly it back to Cam Ranh Bay.

I only pulled Maintenance Alert once during my tour. On that day, a South Vietnamese helicopter got a little too close to a Caribou that was on the ground at Duc Lap. The helicopter blade shredded the rudder but fortunately did not damage the vertical stabilizer so the repair was a relatively simple parts swap and could be completed in a few hours.

Duc Lap was near the Cambodian border southwest of Ban Me Thuot and about 100 miles or so west of CRB.

We left Cam Ranh Bay with a new rudder, several mechanics, and a tripod device that they used to change the rudder in forward areas. We reached Duc Lap around noon and the other crew took our airplane to continue their mission.

While we waited for the aircraft to be repaired, I

struck up a conversation with a US Special Forces advisor who had about 15 Montagnard troops with him. They had been waiting three days for transportation to Nha Trang. I offered to take them when we got the airplane repaired, as Nha Trang was just up the coast from Cam Ranh Bay and we could drop them off on the way home.

He checked with his people and they agreed so we did it.

Special Forces waiting for a ride to Nha Trang.

Wounded Caribou at Duc Lap. Note our airplane departing right behind the damaged rudder.

Caribou on ramp at Duc Lap. Firebase behind.

CRASH AT DALAT/CAMLY

The runway at Dalat/Camly was at elevation 5,040 feet above sea level. It had apparently been constructed by leveling the top of a mountain. On one end of the runway, the terrain sloped very steeply down from the airport elevation. When you flew final on that end of the runway, this slope significantly changed the "picture" that the pilot saw. The effect was to make you feel that you were above the normal glide slope. We were all taught to beware of this effect and be very careful to compensate for it.

In one of the other squadrons, I think the 535th, there was a young officer who was handsome enough to be straight out of central casting. His family had some political connection back in the States and he was therefore the "golden boy" in the Wing. He had very quickly been upgraded to Instructor Pilot and if my memory serves me, assigned to Standardization as a check pilot.

On a flight into Dalat/Camly, he was observing a pilot and co-pilot make the landing. Their cargo was a 5,000-pound rubber fuel bladder. The pilot flying the Caribou evidently fell victim to the corrupted "picture" and ended up crashing the airplane on the runway. As I remember it the Caribou came to rest off the left side of the runway about halfway down. With the fuel bladder on board, there was the obvious hazard of a fire.

The entire crew escaped without significant injury, but the airplane was a total loss. Years later, I became friends with an Army Huey pilot who just happened to be there. He had landed his helicopter and assisted the crew in exiting the aircraft. His description of the incident was much less than flattering for the handsome young Instructor Pilot. Despite his less than stellar performance, I heard later that he was decorated for his actions that day.

A BRIEF MOMENT OF PRIORITY

We returned to Cam Ranh Bay one afternoon with passengers on board that needed to de-plane on the commercial side of the base where the Freedom Birds and other transient aircraft operated. After landing, we taxied over to that side and let them off, engines running.

From there, we had to taxi on the parallel taxiway to the approach end of the runway, turn right to cross the runway to the C-7 side of the base and then taxi on to the Caribou ramp to park for the night. We were done for the day.

For many pilots in the Air Force, particularly those unfamiliar with the C-7, our aircraft was something of a red-haired stepchild. It had been bought originally by the Army. It was ugly. It was a bit underpowered. It was not a JET. Most of us who flew the C-7 learned to love it. You could cross-control the thing (full rudder one way and full aileron the other) and make it sink like a rock and then, in the blink of an eye, neutralize the controls and be flying again. The short-field performance was a sight to behold. As much as we loved it, though, we were well aware that others in the Air Force didn't appreciate those qualities and we learned not to expect priority on anything. It reminded me of my Grandfather's saying, "*Blessed is he that expectith nothing, for he shain't be disappointed.*"

On the other end of the priority spectrum was the

C-5 Galaxy, the newest and largest cargo aircraft in the Air Force inventory. C-5's landed frequently at Cam Ranh Bay and they used the commercial side for unloading and refueling. They never stayed there overnight for fear of being attacked with rockets launched by the VC or NVA from the adjacent mainland. Outside of B-52's that almost never landed in-country, the C-5 had the highest priority of any aircraft. Everyone else worked around their schedule and requirements.

This afternoon there was a C-5 on the ramp preparing to depart, engines already running. As we were traveling down the parallel taxiway toward the approach end to cross, he started taxiing out at 90 degrees to our path to take the runway. Our taxi routes were conflicting.

My first thought was that they would obviously make us hold and clear him to take the runway. He would sit there, run up his engines, complete the checklist, and wait 5-10 minutes for clearance from Air Traffic Control. We would have to wait until he departed, just to cross the runway. We were tired after a long day. It was hot. We were not in the mood to sit there unnecessarily when we were just a minute or so from our home ramp.

To our great surprise, as the two aircraft approached a point where something had to give, Ground Control said, "C-5, HOLD FOR THE C-7."

We couldn't believe our ears! We whooped and hollered like crazy people. Imagine, a C-7 getting taxi priority over a C-5! No matter that it was perfectly logical. This just *does not happen*!

THIEN NGON RE-SUPPLY

On 17 March 1971, we were in Bien Hoa flying the seven-day stage. After completing our assigned sorties for the day, we returned to the ramp at Bien Hoa. The Major in charge of the Bien Hoa Caribou operation drove up to the airplane and asked us if we would volunteer for one more sortie.

The story he told was that they needed to get some ammunition up to Thien Ngon Airfield in the Parrot's Beak area. A nearby base had been overrun the night before and they were worried that Thien Ngon would come under attack that night. I had landed at Thien Ngon many times before and had been flying in and out of the Parrot's Beak area all that day, so we volunteered.

The airlift guys brought us a load of ammunition and we taxied out and took off. Thien Ngon was about 60 miles from Bien Hoa so by the time we got there, daylight was starting to wane. The bigger problem, though, was that there were several aircraft ahead of us, maybe one or two Caribous and a C-123 or two. With one airplane on the ground at a time, this was going to take a while.

When our time finally arrived, it was dark. Caribous typically did not make night landings at unlighted forward fields. The runway was 2,900 feet long, 60 feet wide, with a laterite (dirt) surface, but it was level and had a good off-loading / turnaround ramp area on one end. It was a Type 2 for a Caribou, meaning there was plenty of runway length.

We could see well enough to position ourselves on a downwind and make a final turn. With that much runway, we could afford to be cautious, holding the aircraft off until we were certain we were over the runway. We delayed using our landing lights until a few seconds before touchdown and once on the ground, immediately turned them off. The off-loading and takeoff were similarly uneventful.

Weeks later I was contacted by the Squadron about awarding me the Distinguished Flying Cross for that mission. I felt honored but thought then and have thought since that there was a great deal of distinguished flying going on in Vietnam. Unfortunately, much of it could not be described in detail in an award citation because it reached or passed the limits of the rules we were operating under. For example, the award citation for the DFC I received for that mission says nothing about it being a night landing since Caribous in Vietnam, according to policy, did not do night landings at forward fields.

In 2018, my grandson who was 11 years old at the

time, asked me, "Grandpa. Were you really good at what you did in Vietnam?" This question surprised me because it was totally out of context, so it was probably something he had been thinking about. I told him that he had stumbled onto a really interesting fact about Caribous and similar aircraft in Vietnam.

The pilots of most larger aircraft, and other types of aircraft with more range and payload will typically perform one or two takeoffs and landings in a day. Except for weapons delivery, these are probably the most critical phases of flight where the skills of the pilot are challenged the most. In Caribous, in an 8 or 10-hour day we might make 8, 10, or even 12 takeoffs and landings. We got a lot of practice flying those critical phases of flight and so most of us got really good at it. I think this repetition was particularly important to me because I require multiple repetitions to learn physical skills. Another reason why the Caribou turned out to be a very lucky assignment for me.

THE UNITED STATES OF AMERICA

TO ALL WHO SHALL SEE THESE PRESENTS, GREETING:
THIS IS TO CERTIFY THAT
THE PRESIDENT OF THE UNITED STATES OF AMERICA
AUTHORIZED BY ACT OF CONGRESS JULY 2, 1926
HAS AWARDED

THE DISTINGUISHED FLYING CROSS

TO

FIRST LIEUTENANT ALAN C. GRAVEL

FOR
EXTRAORDINARY ACHIEVEMENT
WHILE PARTICIPATING IN AERIAL FLIGHT

17 March 1971
GIVEN UNDER MY HAND IN THE CITY OF WASHINGTON
THIS 12th DAY OF November 1971

GENERAL, USAF
COMMANDER, SEVENTH AIR FORCE

SECRETARY OF THE AIR FORCE

DFC Certificate

CITATION TO ACCOMPANY THE AWARD OF

THE DISTINGUISHED FLYING CROSS

TO

ALAN C. GRAVEL

First Lieutenant Alan C. Gravel distinguished himself by extraordinary achievement while participating in aerial flight as a C-7A Aircraft Commander at Thien Ngon Airfield, Republic of Vietnam on 17 March 1971. On that date, Lieutenant Gravel flew a combat essential mission in support of Vietnamese artillery unit engaged in activity against the enemy. Lieutenant Gravel resupplied the base on three successive sorties despite adverse weather, hazardous terrain, and a lengthy exposure to hostile ground fire. The professional competence, devotion to duty, and aerial skill displayed by Lieutenant Gravel reflect great credit upon himself and the United States Air Force.

DFC Citation.

Thien Ngon runway.

Thien Ngon ramp pedestrians.

INSTRUCTOR PILOT STORIES

LOW ON GLIDE SLOPE

In the 536th Tactical Airlift Squadron, we had a tall slender Captain whose name was, I believe, Wayne Walker. He was out of C-141's and after a fairly short indoctrination, was assigned to Standardization as a check pilot. I remember that we sometimes called him "Wing Walker," referring to the ground crewman who ensures that a taxiing aircraft's wings are clearing other aircraft and obstructions. One thing I remember about him was that he was an avid poker player.

The last few months of my tour at Cam Ranh Bay, I was fortunate to be given the opportunity to upgrade to Instructor Pilot. On one of my upgrade training flights, I was sitting right seat, an aircraft commander upgrade student was sitting left seat, and Captain Walker was standing between the seats, observing my instruction.

The communication and navigation radios in the Caribou were mounted in a box on rollers that functioned like a drawer. Once the pilots were in their seats, the "drawer" would be pulled out from under the instrument panel so that it would be accessible between the pilots. To exit the seats, the pilots would push the drawer back under the panel and climb down out of the seats. Because the Caribou was originally purchased by the Army, the radios were a strange collection of UHF, VHF, and FM communication radios, and VOR, DME, and ADF navigation radios. They looked old, maybe even obsolete.

The ADF was tuned by a stack of concentric knobs of decreasing size. The stack resembled a miniature wedding cake, with about five or six tiers. The top smallest knob had a long shaft attached that went down into the tuning mechanism of the radio. Frequently, these would not be secured, and you could lift the smallest knob attached to the three-inch long shaft right out of the radio.

Sometime in the middle of the day, the student was making a routine instrument landing in good weather at some base north of Cam Ranh Bay, maybe Qui Nhon or Pleiku. Everything was relaxed and we were mixing small talk in with the check lists and discussion of procedures. In the middle of all this, Captain Walker grabbed the ADF knob, pulled the attached shaft out of the radio like a dipstick and said, "You are two quarts low on glide slope."

Maybe you had to be there, but at that time and place it was funny.

LANDING PROCEDURE AT DAK PEK

All of the airfields in Vietnam were categorized as Type I, II or III. Tan Son Nhut, Bien Hoa, Cam Ranh Bay, and similar major bases were Type III's. Smaller fields with 3,000 - 6,000 feet or so of runway might be Type II's, although the width, surface type, surface condition, and other factors entered into the classification. Type I fields were those with the shortest, narrowest, most difficult runways. The classification would vary for aircraft type—the same runway might be a Type I for a C-123 and a Type II for the Caribou.

We operated into a number of fields that were also used by the C-123's. Only once did I see a C-130 land at a forward field. At Caribou Type I fields, we were generally the only fixed wing cargo aircraft operating there.

Beyond Type I there were Type I Restricted fields. To fly into these fields, the aircraft commander had to have previously been checked out at that particular field by an Instructor Pilot. Typically, they had unique non-standard landing procedures or challenging conditions. Dak Pek was a good example.

The airfield was surrounded by small hills that were very close to the runway. The Tactical Aerodrome Directory says "C-7 wing will clear hill if wheels on runway." With a cross wind this was particularly exciting as the hills would alternately block and concentrate the winds. On both ends of the runway, the approach was over a small river with an abrupt upslope to the end of the runway. The runway was asphalt, 1,500 feet long and 60 feet wide. There were other problems, but you get the idea.

As I remember it, the landing procedure I was taught was as follows:

Fly the downwind above a small valley that was more or less parallel to the runway. From this downwind you only got occasional glimpses between the mountains of the runway off to the side.

When your left wing comes abeam of a certain large dead tree on the hill, turn 45 degrees left between two small hills and begin the descent. You cannot see the runway from here.

When you reach a small river, turn 90 degrees left. You are now 45 degrees from the runway heading and continuing to descend. On this leg you will begin to see the end of the runway.

When you intercept the extended runway centerline, turn the final 45 degrees left and land. In making this final turn, you have just enough time to level the wings and the runway is under you.

During this procedure, you were in a constant descent, hoping to be on glide slope when the runway appeared. The abrupt upslope just off the end of the runway distorted the normal landing "picture" so you had to be careful not to be too high or too low.

Dak Pek runway.

Dak Pek firebase.

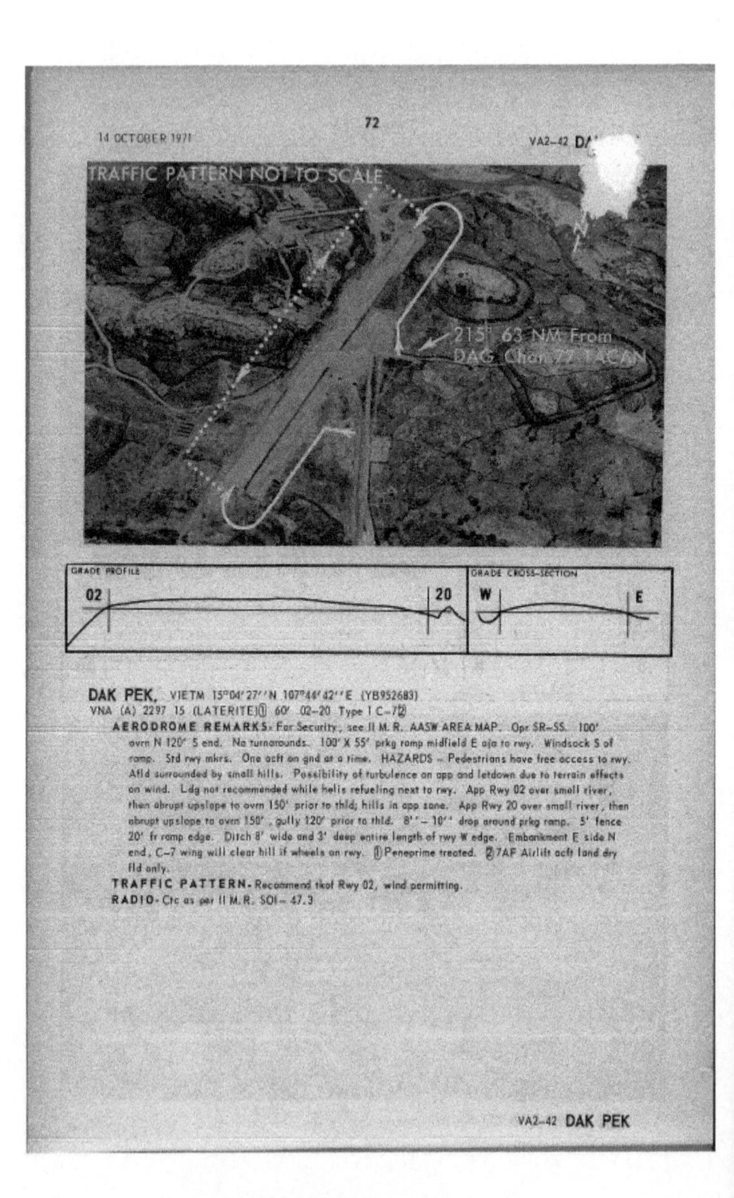

14 OCTOBER 1971 72 VA2–42 DA

TRAFFIC PATTERN NOT TO SCALE

215° 63 NM From DAG Chan 77 TACAN

GRADE PROFILE 02 20

GRADE CROSS-SECTION W E

DAK PEK, VIETM 15°04′27′′N 107°44′42′′E (YB952683)
VNA (A) 2297 15 (LATERITE)(1) 60′ 02–20 Type I C–7(2)

AERODROME REMARKS- For Security, see II M.R. AASW AREA MAP. Opr SR–SS. 100′ ovrn N 120′ S end. No turnarounds. 100′ X 55′ prkg ramp midfield E adj to rwy. Windsock S of ramp. Std rwy mkrs. One acft on gnd at a time. HAZARDS – Pedestrians have free access to rwy. Afld surrounded by small hills. Possibility of turbulence on app and letdown due to terrain effects on wind. Ldg not recommended while helis refueling next to rwy. App Rwy 02 over small river, then abrupt upslope to ovrn 150′ prior to thld; hills in app zone. App Rwy 20 over small river, then abrupt upslope to ovrn 150′, gully 120′ prior to thld. 8′′ – 10′′ drop around prkg ramp. 5′ fence 20′ fr ramp edge. Ditch 8′ wide and 3′ deep entire length of rwy W edge. Embankment E side N end, C–7 wing will clear hill if wheels on rwy. (1) Peneprime treated. (2) 7AF Airlift acft land dry fld only.

TRAFFIC PATTERN- Recommend tkof Rwy 02, wind permitting.

RADIO- Ctc as per II M.R. SOI– 47.3

VA2–42 **DAK PEK**

Dak Pek Aerodrome Directory page.

DAK SEANG

Dak Seang was a Special Forces camp which straddled the road up a valley north of Pleiku. I believe it was manned by Montagnard Special Forces with American Army Green Beret Advisors. In the spring of 1970, the NVA had laid siege to the camp in an attempt to overrun it. During the siege, Caribous flew low altitude parachute extraction missions to re-supply the camp. In the first few days of this effort, several aircraft and a number of crewmen were lost. It was the most deadly single effort by USAF Caribous in the war.

The tactics were adjusted to use single aircraft at night with the support of C-119 illuminator aircraft using their light to briefly illuminate the drop zone right before the drop and close air support to suppress the anti-aircraft fire. Caribous successfully supplied the base for 69 more days without further losses. After a few months of siege, the NVA withdrew and re-supply returned to full-stop landings on the runway adjacent to the camp.

The runway was not particularly short or narrow or surrounded by difficult obstacles. The difficulty factor was that there was a serious "ski-slope" profile to the runway. The low point was where the road crossed the runway. The surface sloped upward a short distance in one direction and steeply up and then down again in the other direction. A Caribou was about 47 feet tall but if one was sitting in the low point, it was not visible from another Caribou on the other end of the runway. We typically observed the rule of one airplane on the ground at a time and we kept engines running during the offload.

The women of the Montagnard families that lived in the camp could often be seen sitting on top of the bunkers. I'm sure the bunkers themselves were like an oven on a hot day and so outside, even in the direct sun, had to be more comfortable.

Dak Seang.

Dak Seang on short final thru co-pilot's windscreen.

CHIANG MAI TO BANGKOK

Late in my tour, the chief of standardization was a Major Fry. He had come from the C-141 upgrade school at Altus AFB in Oklahoma. He was, coincidentally, a man of small stature but as personable and nice an officer as I ever met. At this time, I was a lieutenant Instructor Pilot.

The Caribou Inspect and Repair As Necessary (IRAN) facility was located at Don Muong airfield in Bangkok, Thailand. Caribous were also used in Thailand to fly a scheduled courier or diplomatic run. Each morning a Caribou would depart Bangkok, fly a loop to the north, landing at five or six airfields to drop off and pick up passengers, light cargo, and packages. These flights were manned by crews from Cam Ranh Bay. As I remember it, an aircraft commander would catch a hop to Bangkok, fly for one week as co-pilot and then one week as aircraft commander, and then return to CRB. It was a coveted assignment in that it amounted to an unofficial R&R. The whole arrangement worked well because the need to shuttle crews back and forth for the diplomatic run and the need to shuttle aircraft back and forth for IRAN dovetailed nicely.

I was never lucky enough to be chosen for this duty, but my roommate and best friend was. During his week as Co-Pilot he flew with an Aircraft Commander who had been one of our classmates in C-7 upgrade training at Dyess AFB in Abilene. One day as they were taking off from Chiang Mai, Thailand, the top hatch flew open just as they lifted off the runway. This was not a common occurrence, but it was not unheard of either. Official procedure was to enter the traffic pattern, land, secure the hatch, and then take off again. In practice, it was possible to kick left rudder which would alter the air flow over the fuselage and cause the hatch to close. The flight engineer could then secure the latch.

On that day, the Aircraft Commander elected to simply put the aircraft back down on the runway, intending to secure the hatch, taxi back around, and take off again. With about 5,000 feet of runway ahead of him, this might have been a reasonable decision except that he had already retracted the landing gear. The landing gear lever was on the Pilot's side of the overhead console and pretty much out of the view of the Co-Pilot.

The gear-up landing destroyed both propellers and engines, scraped all the antennas off the belly of the aircraft, and created numerous small and large holes in the fuselage. One propeller tip came through the fuselage just above the head of a Thai passenger. Luckily, there were no injuries.

Understandably, this incident was embarrassing to the Caribou operation in Bangkok and to the US government in Thailand. An all-out effort was launched to get the damaged Caribou out of Chiang Mai as quickly as possible. Major Fry came to my quarters in CRB and asked if I wanted to go with him as his co-pilot to fly the damaged Caribou to Bangkok for repair. He knew that I had an engineering education and thought that might come in handy as we inspected the aircraft to determine if it was flight worthy. I jumped at the chance.

We boarded a C-130 at CRB with two engines, two propellers, mechanics, tools, etc. and flew direct to Chiang Mai the next morning. The mechanics got to work while we went into town. The next morning, both engines and both propellers had been changed and other minor repairs made. My memory is that the mechanics had worked all night on the repairs. Most of the holes had been closed with metallized tape. The bottom hatch just below and behind the cockpit had been seriously damaged. The structure around the hatch seemed fine so they closed the hatch and taped over the whole area, about three feet square, with metallized tape.

We performed a very careful pre-flight and took off to perform a double propeller feathering and engine shutdown/restart in the vicinity of Chiang Mai. The plan was to then land, make any necessary repairs, top off with fuel, and take off again for Bangkok. Everything went so well, we somehow decided (I seriously don't remember how the decision was made) to just strike out for Bangkok. We had no weather radar, only one vhf com radio, and one adf nav radio that had been rigged just to get us to Bangkok.

By this time, it was early afternoon and the typical thunderstorms began to cook up. A large thunderstorm developed right ahead of us on our southerly course direct to Bangkok. We began easing around to the west and the visibility got worse and it got very dark. We diverted so far off our direct course that we wondered if we were over Burma and began to be concerned about fuel. The storm got bigger and the farther west we went the farther it seemed we would have to go. Major Fry suggested that we might better return to Chiang Mai and asked what I thought. I suggested that we press on for another ten minutes and that if I couldn't figure out in that time exactly where we were and be sure we had the fuel for Bangkok, then we should turn back. For about five or six minutes, no one said a word on the intercom. It was as pregnant a silence as I have ever experienced. In the middle of this silence, that three-foot square of tape blew off the bottom hatch, making a booming, whooshing sound that must have drained every drop of blood from my face. I know it had that effect on Major Fry—he was as white as a sheet. After a few seconds we figured out what had happened and had a good laugh.

This whole time, I was doing some serious chart reading, trying desperately to figure out for sure where we were. About the time the ten minutes ran out, I saw a large, concrete hydroelectric dam (like Hoover Dam) off to the west of our track. I figured there could be only one of those in the vicinity, so I quickly found it on the aeronautical chart, and we plotted a course for Bangkok. After a while we finally passed the center of the big storm and started seeing some lighter sky to the south. We raised a Thai air controller at Takhli on the vhf and were able to tune an ADF station in Bangkok. Things were looking good except that we were seriously low on fuel.

We declared a fuel emergency with Don Muang tower and requested a straight-in approach. We decided to leave flaps and gear up until the last minute to conserve fuel. At about 1/4 to 1/2 mile out, we put down some flaps and lowered the landing gear. Two green lights on the mains but no green light on the nose gear. We quickly agreed that we were not going around—it had been landed gear-up once and it could be landed gear-up again. That would be much better than fuel starvation on go-around.

Just before the main gear touched down, the nose gear light came on—must have been some air in the system. Just another routine landing.

P.S. Years later I pulled Chiang Mai up on Google Earth and traced our route down toward Bangkok. Sure enough, there was the dam and, although it was close to Burma, it *was* in Thailand. It's amazing how technology has changed our lives.

Navigation check-point dam.

BANGKOK BSU AND BIBLE STUDY

On one of my trips to Bangkok I found myself with a day off and, for some reason, I was by myself. I had been somewhere and was walking back to the hotel when I saw the sign "Baptist Student Union" on the front of a building about a block from our hotel. Just being curious, I went in the front door. It was empty except for one young man who was busy straightening chairs or something. We struck up a conversation and in the course of it I told him of my experiences in BSU at Tech and my summer missions experiences.

After talking for a while, he said that they were going out to a village outside Bangkok to do a Bible study that night and asked if I would like to go along. I agreed and we set a time to meet back at the BSU center.

At the appointed time, about five young Thai men and I got into a VW bus and started driving through Bangkok. Soon we reached the outskirts and continued driving through a very flat area occupied by rice paddies and divided occasionally by canals. After about 30 minutes of driving, we pulled off to the side just before a bridge over one of the canals. They told me we would wait here until the boat came to get us.

After a while one of those long narrow boats with the motor on the back with a long drive shaft pulled up. The same type as in the James Bond movie. We all got in and off we went. We motored down one canal, turned into another, then another, then another until we must have been five or so miles from where we parked the VW bus. It may as well have been 100 miles.

Eventually we pulled up to a small village on stilts. There was no dry land. The main "street" was an open area of water between two rows of houses on stilts above water. There were intersecting "streets" at right angles. We pulled up at one house, climbed up out of the boat onto the raised porch, and went inside. The BSU guy had brought a felt board with him so he told Bible stories, illustrating them by adding pieces of different colored felt to the board until he had a complete picture. I hadn't seen a felt board used like that since I was a young child. I couldn't understand a word he said in Thai, but I recognized the stories from the felt board picture and so could more or less follow what was going on. We prayed and then the host family brought out some food and we ate.

After about two hours, we said goodbye to the host family and got back in the boat. How the driver got us back to the VW in the dark I could not imagine but soon enough there was the bridge and the VW right back where we had left it.

In the years since, I have second-guessed my wisdom in accepting this invitation, but it all worked out and I had a unique experience that I will always remember.

Bible Study somewhere near Bangkok.

BANGKOK SHOPPING AND DINING

I was in Bangkok three times during my tour. Since the Caribou IRAN (Inspect and Repair As Necessary) facility was in Bangkok, the 483rd wing had permanent party there to manage shuttling the airplanes in and out and perform oversight of the contractor doing the IRAN work.

The permanent party staff had developed relationships with various vendors who they could recommend if you wanted to purchase Thai goods such as wooden dinnerware, ceramic goods, brass goods, jewelry, etc. I bought some jewelry there for Sheri and a few other things.

They also had several favorite restaurants where we would eat as a group. One night that I was there a group of about eight or ten Caribou personnel were to go out to eat. They recommended a place and I went along.

When we entered the front door, the first thing we saw was a small bleacher area with 15-20 young girls, maybe 12-13 years old, sitting there watching some cartoon show on television. The setup they had was that you would choose one of these girls and from that point on you would not have to lift a hand to do anything for yourself. They showed you where to sit (on the floor), tucked your napkin in at your neck, served your plate from the dishes spread out in front of you, fed you, wiped your chin if something went astray, and held your goblet for you to drink. It was considered an insult if you did anything for yourself.

This was a bit bizarre for me and made me uncomfortable, but I went along with it as best I could and filed it away as one of the stranger experiences I would have in Southeast Asia. I tipped my little girl generously, reasoning that if she is doing this to earn money, she must *really* need it.

R & R IN HONG KONG

During the year in-country we were allowed two weeks of Rest and Relaxation. On the first trip, early in 1971, I met Sheri in Hawaii and we spent the week together near Waikiki Beach.

For the second trip in the summer of 1971, David and I went to Hong Kong together. There was a large Navy Exchange and extensive civilian shopping opportunities in that area, so we arrived with lots of cash and requests from many squadron mates for things they wanted us to buy for them, particularly stereo equipment and cameras. We shopped, ate some good (and some not-so-good) food, and saw the sights in Hong Kong.

While in Bangkok and the Philippines, I enjoyed buying jewelry for Sheri. I had sent her a Pacific Exchange catalog asking her to identify some things she would like to have. She didn't give me any help except to say that she really didn't like jade.

On the last day before our return to Vietnam, we had been to Kowloon island on the ferry and were returning to our hotel in the late afternoon. We were walking down the sidewalk of a very urban area with high rise buildings. When we passed a small jewelry store on the ground floor of a multi-story building, I asked David to wait while I went in and looked around.

After a few minutes of looking in the display cases, I spotted a jade and gold ring that I thought was very pretty. Remembering Sheri's comment about jade and knowing that I only had $80 in my pocket to make it back to Vietnam, I tried not to be too interested. The Chinese lady about 65 years old behind the counter was an experienced sales person, and did not miss the cues I was evidently showing.

She said, "You see something you like?"

I replied, "Yes, but I don't have enough money for it."

The ring was priced in Hong Kong dollars, equal to about everything I had. We knew that generally you could buy things for about 3/4 of the marked price, but that was still too much.

She said, "You like something, you make me a fair offer, I will sell it to you."

I said, "No. Anything I can afford to offer would be an insult to your merchandise."

She was persistent so I finally identified the ring I was interested in and she took it from the case and let me look at it up close. I was sold but still didn't have enough money. She kept insisting I make an offer and I kept resisting until David finally said, "Make her an offer. Tell her $10 and let's get out of here."

I finally said, "OK. I can pay $40 US, but I don't expect you to sell it for that."

She looked very carefully at the ring. Her husband came from behind the counter on the other side of the store, got out his eye piece and took a long look at the ring. He finally nodded "yes" to her.

Then she said, "It's 6:25 pm. We close at 6:30. Mama is upstairs waiting for dinner. We make this last sale of the day, *not for profit, but for luck.*"

In my life, before and after that, I have been extraordinarily lucky many, many times. I am so profoundly aware of this good fortune that, as I write this and when I think about that Chinese woman's words, I get goose bumps.

The irony is that of all the rings that I bought for Sheri in Southeast Asia (about five or six), that jade ring was the one that she wore many times more than all of the others combined. She loved it.

UDORN AIR AMERICA CARIBOU

In the summer of 1971, I was given the assignment of returning a Caribou from Udorn, Thailand to Cam Ranh Bay. We were told that this Caribou had been on "loan" from the US Air Force to Air America for several years and that they no longer needed it. Air America operated Caribous in country and we frequently would see them. Their Caribous did not have the camouflage paint scheme ours did, they were shiny aluminum. Another lieutenant aircraft commander from our squadron was assigned to accompany me as co-pilot, along with a flight engineer. We flew a Caribou to Don Muang in Bangkok, delivering it for its periodic going-over at the IRAN (Inspect and Repair As Necessary) facility there.

The next morning we boarded a C-130 for the ride to Udorn. On arrival, we checked in with the Air America people. They told us the aircraft had been repainted in standard Air Force camouflage and it would be ready the next morning. We went into town and checked into the hotel used by USAF transient crews.

The following day we returned to the Air America facility at the airfield and performed a fairly thorough pre-flight on the aircraft. Since the aircraft had not flown much in the last few months, it was not surprising that one of the magneto ignitions on one of the engines was inoperative. The R-2000 engines on the Caribou had dual redundant magnetos and you could fly with only one system operative, but we generally did not take off unless both systems were operating. We taxied from the run-up pad back to the Air America facility and informed them of the problem. They told us that their mechanics would not be able to get to the problem until later in the day, so we returned to town and checked back into the hotel.

We repeated this the next morning and again there was some problem with the airplane that prevented us taking off. We sensed that the Air America people thought we were just looking for some problem, or worse yet, sabotaging the airplane to extend our stay in Thailand. They were ready to be rid of the airplane and seemed aggravated with us for not just taking it as-is.

On the third day, we finally had a clean airplane and flew down to Bangkok. Air America was understandably concerned that the transfer of ownership of the airplane be documented so before we left Udorn, they made me sign personally for the aircraft.

The next day we flew that airplane from Bangkok to CRB. We arrived late, about dark as I remember it, left the airplane with the ramp personnel, and went to our quarters. The Air Force never officially signed the airplane back over so I guess somewhere in the world there might be a Caribou or the remains of one that still officially belongs to me.

GIA VUC

Toward the end of my tour, while flying sorties out of Phu Cat, I was assigned a load for Gia Vuc. I had never been there before and have never met any other Caribou pilot who had. It is located 80 miles or so south of Da Nang in a north-south valley. The area was spectacularly beautiful and were it not for the well-rusted concertina wire, you would never have known there was a war going on.

When we landed, it was very quiet and peaceful, so we shut down for the unloading and walked into the village to take a few pictures. As was usual in those remote locations, the children gathered around and this time a young mother approached us with her children. They were very curious about the co-pilot's camera.

On the ground at Gia Vuc.

Overview of Gia Vuc.

FINI FLIGHT

Although we were roommates for the year, David and I never flew together. We were both co-pilots at the same time and then aircraft commanders at the same time so there wasn't the opportunity. For your final flight, however, it was customary for the squadron to accommodate requests for unusual crew selections. We asked for and were given a flight together. A young college-educated clerk from the squadron asked if he could fly the day with us and was given permission.

We would alternate positions at co-pilot and a/c

while performing a standard mission. I don't remember all we did but I know we flew down to Bien Hoa and flew sorties out of there, one of which involved transporting a seemingly vicious US Army K-nine in an aluminum crate/cage. At one point during the day, we landed at Dalat / Lien Khoung.

My thinking about our final flight and the special things we might try to do that day was tempered by a recent event that had occurred on a fini flight. The C-123's were based in Phan Rang, about 30 miles down the coast from Cam Ranh Bay. The story we heard was that four or five C-123 pilots who had trained together in the States before going to Vietnam had requested that they be allowed to fly their fini flight together. The group was a mixture of young lieutenants like us and several higher-ranking pilots, possibly a major and a lieutenant colonel. At the end of the day, in a spirit of celebration, they made an unauthorized low pass on the control tower at Phan Rang, intending to pull up into a downwind and land for the final time. Somehow in the pull up something happened, perhaps they stalled the airplane. The aircraft crashed and all souls on board were killed. Tragic deaths, and particularly so since they had made it through the year and the final event was so unnecessary.

On our fini flight we didn't do anything stupid, but we did have some fun. Early in our year, a guy who was leaving to go back to the States had given David an AK-47 and some ammunition. We had never done anything with it. That day when returning to Cam Ranh Bay feet wet, we opened the rear cargo door and shot all the ammunition out into the air over the ocean. At one point I un-zipped my flight suit and pushed it down out of camera range for a picture of me "flying naked." We had a fun day, mostly doing the things we had done all year.

When we landed, as was tradition, we were met by the Officer of the Day. He brought champagne and we had a few drinks out on the tarmac around the aircraft. The custom was that when uncorking the champagne, you would try to shoot the cork into the engine augmenter tubes (exhaust pipes) on the top of the wing.

After 30 minutes or so of celebration, we headed up to the squadron office to turn in our gear. The young clerk who had flown with us that day bought and opened a beer for me. Not knowing anything about alcohol consumption I drank it on top of the Champagne. By the time we got to the hootch, I was not feeling too good. I threw up, took a shower, threw up again, and finally tried going to bed. I don't think I have ever been that sick, before or since. David and some other guys came by later and asked if I wanted to go to eat and I had to decline. My year in Caribous was over and I was in no shape to celebrate.

Me flying naked.

David and I after Fini Flight.

LAST NIGHT IN CAM RANH BAY

The day I arrived in Cam Ranh Bay some rockets hit the other side of the base, closer to the mainland. One rocket just happened to hit the Navy Officer's Club and killed a Navy officer who was having breakfast before boarding his Freedom Bird to go home. He had survived a full year in Vietnam and was killed the morning he was to leave. This story had haunted me all year and there was a final event that would bring it to mind one more time in a very dramatic way.

In our remodeled hootch, we had bunk beds to save space. David slept on the bottom and I slept on the top with less than a foot of clearance between my face and the ceiling.

We were both scheduled to leave on the same day, David at something like 9:00 am and me at around 11:00 am. We had already sent stereo gear and other non-essential things home in "hold baggage." The night before, we packed our clothes and the few other things that remained. As was usual, we each got in our bed and had a short conversation in the dark before falling to sleep.

Sometime in the very early morning, just before dawn, we woke to a tremendous series of explosions. The plywood ceiling of the hootch was blown loose and fell on top of me in the top bunk, along with the dirt and insulation that had been in the attic. I scrambled out from under it and got on the floor under the bottom bunk alongside David who had beaten me there. We had a quick conversation about what might be happening. Before we could come to any conclusion, one of our squadron mates came into the room and yelled, "Hey, ya'll are missing the fireworks show."

We sheepishly came out from under the bed and went outside. A large group of guys were gathered out there with some on the roof of the Quonset huts where they had a better view. The enemy sappers had gotten into the bomb dump and set off some explosions which, in turn, ignited some of the munitions stored there. Among those were the 15,000 pound "Daisy-Cutter" bombs that were dropped by C-130's to clear landing zones for helicopters. The explosions were still going off and in the early dawn, you could see the pressure waves from the explosions radiating out of the bomb dump area. It was quite a sight and sound experience and you could even feel the pressure waves.

As David's Freedom Bird approached for landing, the enemy dropped several rockets in on the runway. They had to go-around and wait for things to calm down. Then we were told they had to go to Saigon for fuel. In the meantime, my Freedom Bird landed, loaded up and took off. So, David who was supposed to leave before me ended up leaving several hours after.

As my year in-country came to an end it seemed like we had been doing this forever and at the same time it seemed like we had only arrived in-country last week. To express this contradiction in the sense of the passage of time, we would say, "The days are long and the weeks are short." I can't logically explain that expression, but it seems to describe the feeling you get when the days are packed with activity. So many events happen in one day that you can't absorb the meaning of one before the next is happening. The long string of events makes the day seem extended, but you are so busy that time flies by and the year is over before you know it. In a war zone I suppose that is merciful.

KC-135 STRATOTANKER

CASTLE TO CLARK

KC-135 TANKER TRAINING AT CASTLE AFB

When we returned from Vietnam in September 1971, I took a month of leave for Sheri and Alan and I to travel around to see family and friends. When we left Louisiana for tanker training in California, we made several stops along the way, visiting the Painted Desert, the Petrified Forest and several other spots. We spent one night at the Lodge on the South Rim of the Grand Canyon. I got up early the next morning and drove to a good spot to get a picture of the sunrise over the Canyon. Sheri and Alan stayed at the room which had large windows on the back, facing a wooded area. The clouds and the fog turned my picture-taking into a waste of time, but several deer came right up to the back window of the room, so Sheri and Alan had a memorable experience.

When we arrived in Merced, CA, we found an efficiency apartment at the Slumber Motel. This was an old-style western motel and they had combined two rooms into one to create efficiency apartments. Again, David and his wife were in another unit in the same place.

The training lasted four months or so, basically through the winter of 1971-1972. In the San Joaquin valley, fog is a real problem at that time of the year. It was not unusual to have wrecks on the freeways involving 100 or so cars. To improve safety on the Interstate highway, the Highway Patrol would publish the time that the Highway Patrol would pass a certain exit traveling exactly 55 mph. Drivers were asked to wait at the entrance ramp for the "train" of cars to pass that were following him and then join in.

We drove up to San Simeon to see the Hearst Castle one weekend. The fog was so thick that we couldn't see anything of the castle and grounds.

One night we decided to go to the drive-in movie. We got Alan dressed for the night, expecting him to go to sleep. The movie was "Summer of '42" and after all the cartoons were shown and I got our drinks and popcorn, the movie started. About 15 minutes into it, the fog suddenly just moved in and the image being projected no longer reached the screen. You could see the image in the fog, almost like a hologram, but not very well. After a few minutes, they made an announcement that it was over for the night and directed everyone to come back by the gate to receive a refund.

During our months in Merced, we made multiple

weekend trips to San Francisco, Carmel, and the Monterrey Peninsula. We also traveled to Yosemite several times, once for Christmas. We took my Akai reel-to-reel tape deck to Los Angeles for repair and weeks later went back down there to pick it up. On one of those trips we visited Disneyland.

With the fog, instrument training in the KC-135 was very realistic. On any day there might be four or five tankers practicing Ground Controlled Approaches (GCAs) at one time. When you lined up on final approach, you could often see a tanker a few miles ahead and one or two on the downwind leg of the pattern. At some point in the final descent, the top of the fog was so flat and the loss of visibility so abrupt that you would have the sensation of sinking into a pool of water. On climb-out, a similar sensation would hit you when from total obscuration you would suddenly have total clarity.

All in all, tanker training was reasonably pleasant. We were happy to be back in the States with our families and there was lots to do and see in California. All good things come to an end and we left for our permanent assignments around the first of March 1972. I went to Barksdale in Shreveport, LA, to join the 913th Air Refueling Squadron of the 2nd Bomb Wing. David's orders took him to McCoy AFB in Orlando to fly air refueling missions supporting the SR-71 Blackbird Reconnaissance Plane.

BARKSDALE TO CLARK

When we finished tanker school at Castle AFB in Merced, CA, we reported to Barksdale AFB in Bossier City, LA. We moved our mobile home from Alexandria, LA, where Sheri and Alan W. had lived while I was in Vietnam in Caribous to a nice mobile home park in Princeton, LA, a few miles east of Barksdale. We settled in and started training flights about the first of March 1972. I was a senior 1st Lieutenant with almost 1,000 hours of combat time in Caribous in Vietnam, but I had only the upgrade training hours in the tanker, so I started out as a co-pilot. I was assigned to a crew led by an Aircraft Commander who had come to tankers right out of pilot training, accumulated the minimum required 500 hours in the tanker, and had just recently been upgraded to Aircraft Commander.

Early in May, virtually the entire squadron was deployed to Clark AB in the Philippines to fly missions into Vietnam as part of the US response to the 1972 Easter offensive when the North Vietnamese tried to overrun South Vietnam. Unlike my other experience in Caribous, we deployed as a unit. The morning we left Barksdale was a very dramatic scene, with wives and children crying and saying their goodbyes and the aircrew and support personnel preparing to embark on a long flight across the Pacific while at the same time bidding farewell to their families, and having no idea how long they would be gone.

We crossed the Pacific in two legs, Barksdale to Hickam in Hawaii and Hickam to Clark in the PI. We flew three ship trail formation, with each aircraft a mile behind and 1,000 feet above the aircraft ahead. As a junior crew, we were third on both legs. In and out of the weather, this arrangement concerned me enough that I never left my seat for 16 hours on the Hickam to Clark leg. I remember we arrived at Clark with 12,000 pounds of fuel remaining, likely not enough to make any reasonable alternate landing site.

While at Clark, we lived in small mobile homes, one crew to each trailer. Something like 60-80 of these were arranged in a grid with asphalt streets between them. We had a kitchen and ate lots of TV dinners. On days off we could play tennis, go to the library, take in a movie, and visit the BX or Commissary.

On that initial deployment, we took 11 airplanes and 14 crews. The whole operation was commanded by a major. Before we left Clark in late summer, there were something like 40 airplanes, 60 crews, and we had three or four full bird colonels running the operation. It was much more enjoyable at the outset when we were flying more and had fewer people telling us when to go to the latrine.

At that point we were launching an airplane every hour, 24 hours a day. We took off at 255,000 pounds, which was considerably lighter than our nuclear war (Emergency War Order) weight back in the States, but

Clark had a pressure altitude around 1,400 feet so the KC-135 seemed over-loaded and under-powered, a lot like they did on EWO launches in the States.

Toward the end of the summer, "Notices" and "Directives" were appearing on the bulletin boards at an unbelievable pace. Most crews resented this micro-management as most of these directives were conceived and promulgated by senior officers who were not actually flying the missions. Someone (I never knew who and didn't want to know) started stamping these documents with a "Bullshit Disapproved" rubber stamp that had, in addition to the words, the image of a bull defecating. Several notices appeared warning that person to cease and desist but, predictably, those got promptly stamped as well.

KC-135 refueling F-105's (stock photo).

MEDICAL SERVICE CORPS OFFICER FROM UT

Back at the University of Texas, when I learned that the draft board was going to limit the 2S student deferment for graduate school to one year, I started looking for ways to satisfy my military requirement. My undergraduate degree was in Civil Engineering and my graduate degree, when I eventually got it, would be in Environmental Health Engineering. These were perfect preparation for a USAF Medical Service Corps officer. They supervised the water treatment, waste treatment, vector disease control, and other public health functions at Air Force bases all over the world.

Under that program, they offered direct commissions, much like they did for medical doctors. I looked into it but got nowhere. Some earlier graduates of the program at the University of Texas had succeeded but it is likely the Air Force had many more qualified applicants than they had positions.

Fast forward three years and I am in the Base Exchange at Clark AB. I recognized this fellow that had been ahead of me in the program at UT and had succeeded in getting a commission into the Medical Service Corps. We talked and caught up a bit. He was Permanent Party at Clark and living in a single-family home in base housing. He invited me to come out to the house for dinner one evening and we had a nice visit. I had not known his wife at UT, but she was very nice, and it was great to get a home-cooked meal after all the TV dinners.

The town just outside the gate to Clark was Angeles City. It was not closed to us, but we were warned to only go in groups and preferably in daylight hours only. I went with a group once or twice to eat at a Mexican restaurant but that was the extent of my off-base meals. That made the home-cooked meal at the home of my UT friend even more of a treat.

Clark AB was a huge facility. Their home was in a

part of the base that I had never seen and like other parts of the base, it was spectacularly beautiful. The parade grounds on the main base was a huge square of beautiful grass surrounded by mature live oak trees that rivaled anything you could find in New Orleans. The large houses across the surrounding street were built 6-8 feet off the ground with wide staircases up to screened porches. After the Vietnam War, the relationship between the United States and the Philippines became strained so there was doubt as to whether we would be able to maintain Clark AB. The eruption of Mount Pinatubo made that a moot point as several feet of ash covered the base, causing fires that further damaged what had not already been vandalized.

I have heard recently that the base has been reopened as a commercial airport. Perhaps in time all that beauty will be restored.

ENGINE FAILURE ON TAKEOFF

Each day one crew would be designated as a "maintenance alert" crew. Your job was to pre-flight an airplane and taxi it down to the approach end of the runway, pull into the run-up area off to the side of the taxiway, and shutdown. We could plug a headset into the airplane and monitor the ground and tower frequencies while sitting outside on the tarmac. Most of the time, you would just wait all day there and never have anything to do. If one of the crews launching had a problem with their airplane that could not be quickly repaired, or if they aborted on takeoff, or if they had to return to base in the first few minutes of their flight, you would launch and take their mission.

On our day on maintenance alert, we had waited several hours and watched one tanker after another take the runway and takeoff. One of the Barksdale crews taxied by and waved and we waved back. When they were well into their takeoff roll, over tower frequency, we heard “Abort, Abort, Abort”. We jumped up, got into the airplane, started the engines, took the runway and got clearance to takeoff. Just as we lifted the nose gear off the runway at rotation speed, the outboard engine on the left side exploded. This was the upwind, outboard engine which was the most critical one for maintaining directional control. This failure at this phase of the flight was the “skull and crossbones” scenario for KC-135 tankers. They tended to roll over on their back which would obviously be fatal when close to the ground.

The crew that had aborted was taxiing back to the ramp on the parallel taxiway and had a front row seat to the explosion. They said flames shot 30-40 feet out the front of the engine and 60-80 feet out the back. To us it felt like someone had hit us in the left shoulder with a sledgehammer.

However, our training kicked in perfectly. The A/C

and I both locked our legs on the right rudder, and we pushed the other three engines up to maximum power. We leveled off about 500-600 feet and turned about 30 degrees left to avoid the mountains ahead and stay over the rice paddies. I got the landing gear and flaps up and started dumping fuel to reduce our weight. We declared an emergency and asked air traffic control for radar vectors to return to Clark. We dumped about 100,000 pounds of fuel until the airplane could climb and be reasonably responsive to the power we had available.

The throttle setting for takeoffs in the KC-135 were made to a certain Engine Pressure Ratio (EPR), the ratio of the pressure in the back of the engine to the pressure in the front. This number was typically something like 3.94 and there was a gauge on the instrument panel that displayed the EPR for each engine. This ratio was calculated during flight planning and was affected by temperature, pressure altitude, etc. The desired setting was definitely not full throttle, as that would result in an EPR that would damage the engines. You were only supposed to use full throttle during dire emergencies such as engine failure on takeoff and then only for the duration that was absolutely necessary.

As soon as things were under control, and in accordance with the emergency procedure for this situation, I told the A/C that I was pulling the throttles back to a normal power setting for climb-out. He stopped me, saying, "No, No, we need the power." I tried several more times and each time he stopped me. He insisted on maintaining the maximum power setting until we leveled off at 10,000 feet. When we pulled the throttles back through about 82% rpm, the whole airplane shuddered and we knew we had *another* problem.

While level at 10,000 feet and following the radar vectors back around to land, we increased the power of each engine separately and determined that it was the other engine on the left side, #2, that was vibrating badly at 82% rpm. We set that engine below the vibration level, set the #4 outboard engine on the other side to idle, and used #3, the inboard engine on the right side to make all power adjustments.

We made an uneventful, albeit tense, approach and landing back at Clark. They had scrambled the fire trucks and the ambulances, and every senior officer who had a blinking light on his car was out on the tarmac to watch us bring it in. We pulled into the ramp and shut down and were immediately surrounded by a crowd. The Navigator and I were so angry at the A/C that we decided we better not say *anything* for fear of ending his Air Force career.

While the crowd was standing around and we were trying to regain our composure, the maintenance personnel opened the cowling on the #1 engine that had exploded. About a dozen turbine blades fell out on the pavement. I picked one up and for years have kept it in my office desk drawer as a reminder of my good luck.

Later we were told that the main thrust bearing on the engine had failed and the rotating parts had moved longitudinally into the stationary parts, causing all hell to break loose inside the engine. I understood that those engine housings are supposed to contain disintegrating parts so how the turbine blades got outside the engine is a mystery. We later learned that the other engine on that same side that had been vibrating was found to be failing in the same way, just not catastrophically.

TWO STORIES WITH AN INTERSECTION

We deployed TDY to Clark AB in response to North Vietnam's invasion in the spring of 1972, normally referred to as the Easter Offensive. The permanent occupant of Clark AB at that time was a wing of F-4's. For the same reason we went to Clark, they deployed to Thailand. I seem to remember that it was Takhli, but it might have been Udorn or Ubon or some other base in northern Thailand. As a result, Clark was occupied that summer by the wives and children of the F-4 pilots and a population of tanker crews, all men, without their wives and families. For the most part this caused no problems, but there were a few circumstances that developed.

Our Navigator's marriage was falling apart. I don't think I knew at the time what the exact status was but suffice it to say that he was interested in female companionship. How he met this lady I do not know, but he developed a friendship with the wife of one of the F-4 pilots. She had heard that her pilot husband was co-habitating with a Thai woman and felt justified in having a social life of her own while he was gone. One of the things that she and the Nav did was play tennis. By that time the tanker operation had grown to about 60 crews flying 24 missions a day, so we had time off that we had not had earlier.

This lady had a good friend whose F-4 pilot husband had been shot down over North Vietnam and was MIA. She didn't know whether to return to the States to wait for the outcome or to stay at Clark. In an effort to take her mind off of her troubles, the Nav's friend asked him to find another tennis player so they could invite her to join them for mixed doubles. The Nav asked me and I agreed.

We played tennis a number of times and followed that with lunch at the Officer's Club a few times. There was no significant interaction between me and this other woman. I tried to be extra respectful of her situation and she seemed to appreciate getting out of the house but that was the limit of it.

The major maintenance functions of the F-4 wing had remained at Clark so on a regular basis, the airplanes had to be cycled through Clark for major repairs and inspections. They could not fly non-stop from Clark to their TDY base so were compelled to make a fuel stop in Da Nang. Aside from the risks of being on the ground at Da Nang, this was an inconvenience to them and added several hours to the trip.

Simultaneous with their trips back and forth between Clark and Da Nang, we had a tanker departing Clark every hour of the day to Purple Anchor just north of Da Nang. It didn't take long for the leadership of the two groups to work out a deal to coordinate the takeoff times and have their west bound returning F-4's top off from one of our tankers just before coasting in. That eliminated the fuel stop in Da Nang and shortened their trip significantly.

We took off one day for Vietnam and had been told to expect an F-4 to pull up behind us for a sip of fuel. When the F-4 arrived, the boom operator said to the Nav on the intercom, "You're not going to believe who this is." The Nav's female friend's husband.

LEFT 90 RIGHT 270 TO LAND OPPOSITE DIRECTION

One night in the early summer of 1972, we were returning to Clark from Vietnam. As we approached the base, the Navigator could see on his radar scope a large thunderstorm about 20 miles or so off the approach end tracking straight at the runway on more or less the same compass heading. We made our downwind and turned a 10-mile final between the storm and the runway, intending to land in front of it.

During the final approach the Navigator was calling out our altitude relative to the glideslope. This was all perfectly normal for this situation. The A/C was flying the airplane. A mile or so out the Navigator called "100 feet below glideslope". The A/C did not react appropriately. A few seconds later, "150 feet below". Still no reaction. A few more seconds, "200 feet below". By this time we are one-half mile out and this is not looking good. I looked over at the A/C and it was as if he was in a trance, catatonic or something. I said "Do something _____!" Still nothing. At that point, I shoved the throttles up, pulled back on the yoke, and called "Go Around" on the intercom and to tower.

The A/C complied with my "Go Around" and continued flying the airplane, climbing slowly back up toward traffic pattern altitude. I asked, "What are you going to do?" and there was no answer. By now the storm was nearing the approach end of the runway and we needed to get the airplane on the ground quickly. We were over the departure end of the runway and getting farther away from where we could land every second.

Somewhere from the deep crevices of my brain, I have no idea how or why, came my solution to the problem. I asked the tower for a "Left 90, Right 270 to land opposite direction." To his credit, the tower controller immediately knew what I was asking for, although extremely unconventional, and he "Approved" without hesitation. I told the A/C to turn left and he did. We drove away from the runway a few seconds and I told him to turn 270 degrees back to the runway and he did. We rolled wings level on a very short final and landed. As soon as the main landing gear touched and before the nose gear touched, we hit extremely hard rain and could see nothing out the front windscreen. He watched the white line on his side of the runway and I watched the white line on my side as we rolled out.

The A/C never said a thing about this. Not "you were insubordinate," not "I could have made the landing," not "kiss my ass," not "thanks for saving my ass." The Navigator and I determined that we would ask for a crew change at the first opportunity.

RED ALERT IN GUAM

At some point in our TDY at Clark, we were selected to go to Guam to sit what was called "Red Alert". Basically, several tankers were parked on the ramp and kept ready in case a B-52 needed unscheduled re-fueling to get back to Guam after a bombing mission over Vietnam. The tanker crews were assigned quarters, but to speed up the launch in the event one was needed, we lived on the airplane while on alert duty. It was hot and uncomfortable and boring, but the B-52 commanders set the rules and the tanker crews were obliged to live by them.

We were there for a week and never had to launch. I don't think we ever even started engines. We did, however get a glimpse of how the B-52 crews lived on Guam and that made us very grateful for our circumstances at Clark. Nothing like seeing someone less fortunate to give you an appreciation for what you have.

TYPHOON TOO LATE TO EVACUATE

At some point in the summer of 1972, a typhoon was approaching Clark AB. Our crew was not there, so it must have been the week we were in Guam for Red Alert. The story that was told is that the Commanders of the 4102nd tanker squadron delayed making the decision to evacuate the airplanes until the winds were already out of limits for takeoff. Clark only had one runway so there was no possibility of using the crosswind runway to avoid the problem.

The solution to the problem was clever. The airplanes were fueled-up to make them as heavy as possible. Several crews were assigned to each KC-135 on the ramp. Engines were started and the crews monitored Ground Control frequency for instructions. Throughout the duration of the storm, as the wind direction changed,

Ground Control would command a new heading and every crew would taxi their airplane to turn to that heading. With the weight maximized, flaps up, and headed into the wind, the airplanes were not susceptible to being blown around by the wind.

CCK TO HOME

CLARK TO CCK

In the first 28 days of June in 1972, it rained 98 inches at Clark AB. Up in Bagio in the mountains, it rained 176 inches in that same period. We were leaving Clark every hour with about 150,000 pounds of JP4 jet fuel which was normally supplied to Clark by pipeline from the Subic Bay Naval Base about 50 miles to the south.

The excessive rains washed out the pipeline to Subic. For a while they tried to maintain Clark's stock of fuel by truck but that was not possible. The decision was made to move the whole operation up to CCK in Taiwan. We were in Southeast Asia on temporary duty (TDY). USAF regulations would not allow a TDY to extend longer than six months so at something less than six months they would rotate you home for 28 days and then they could deploy you again on another TDY. The Barksdale crews had been rotating home for several months and our crew was in the last group to go home.

In early August of 1972, most of the tanker operation moved up to CCK but we only had a few days left before going home so we stayed at Clark until rotating home.

Earlier in the war, tankers had been stationed at CCK. There was a female Chinese bar owner (or madam?) that had catered to the tanker crews during that time. The story was that she had closed her establishment and moved elsewhere when the tankers moved out. Despite the fact that the move from Clark to CCK was supposed to be Top Secret, by the time our operation returned to CCK, she had already come back and set up shop to once again take care of the tanker crews.

In that era there was a Stephen Stills song "Love the one you're with" that was very popular with the bar girls. "If you can't be with the one you love, honey, love the one you're with" seemed to be a perfect message to promote their interests.

Almost exactly in the middle of my 28 days at home, our twins Mark and Ryan were born at Barksdale. I never knew if my squadron commander intentionally scheduled my rotation that way or if it was just a lucky coincidence.

The twins were a surprise. Sheri knew that this pregnancy had been very different from her first so we had concluded that it might be a girl this time. Instead it was twin boys. In the years after, we often wondered if we would have figured it out had I been home for the last five months of the pregnancy rather than in Southeast Asia.

Alan and his boys.

CREW CHANGE

Because of the difficulties we had with the A/C, the Navigator and I decided that we would ask for a crew change while we were back in Barksdale for the 28-day rotation. We went to see the squadron commander and made that request. His answer was that they were not allowing any crew changes in the middle of this "emergency deployment" and that we would just have to find a way to work with the guy. I told him that he may as well just prepare the papers to court-martial me then because I was not going back with that guy as my A/C. At that point he modified his position to say that they would only make a crew change if there was a "safety of flight" issue. I asked him how many stories he wanted to hear. His answer was, in these exact words, "one good one will do." So we told him one.

The story was apparently convincing because he agreed to the change. Looking back, I think he decided to assign us to the most bad-tempered A/C in the squadron and teach us a lesson. We would be sorry we ever bothered him with this problem.

The new A/C had been a fighter pilot, flying F-105's, and had been transferred into Strategic Air Command to fly tankers. His wife was divorcing him. Those two reasons were enough to make him mad at the world. He was gruff and short with his crew and intolerant of anything he thought was a screw-up. He was well known for chewing up young co-pilots and spitting them out before breakfast.

The Navigator and I flew with this A/C for the second TDY to CCK and then back in Barksdale for the 18 months I had left in the Air Force. It was the most professional, most skilled, most fun crew I ever flew with in tankers. We worked together as a team. Each crewmember could do his job and at least one other. The A/C demanded excellent performance on our part but he had the same expectation for himself.

Since the Navigator and the A/C were going through divorces, they enjoyed the night life at CCK, maybe a little too much. We lived downtown so a bus would pick us up a couple hours before our takeoff time. Frequently they would return to the hotel just before the bus arrived. We would ride the 30 minutes or so out to the base, get our briefing, pre-flight the airplane, and go to the flight line snack bar for steak and eggs.

Right after takeoff, the A/C would turn to me and ask, "Do you need me?" He and the Navigator would go to the cargo/passenger part of the airplane behind the cockpit and sleep. The Boom Operator would navigate, and I would fly the airplane.

As we approached Vietnam, I would tune to the Da Nang TACAN (navigation radio). When we locked on with a radial and distance from the station, the Boom Operator and I would "score" his navigation based on how close his calculated position was to the TACAN position. Considering all he could do was "dead reckoning," he was actually very good at it.

From Clark to Vietnam was about two hours. From CCK to Vietnam was an additional 45 minutes or so. We would usually fly over, set up in "purple anchor" north of Da Nang over the South China Sea, refuel and/or be available for refueling until we reached a certain "bingo" fuel and then fly back. Missions ranged from a minimum of about eight hours to a maximum of about 16 hours.

Purple Anchor was a 60-mile long track, from 30 miles north of Da Nang TACAN to 90 miles north along a radial that was more or less due north. We would fly out the radial to 90 miles, turn around, fly back in to 30 miles, and turn around again. We would repeat this until "chicks" (airplanes needing fuel) called on the radio then we would arrange a rendezvous with them.

There were several other refueling tracks around Da Nang, Green Anchor, and Silver Anchor if my memory serves me. They were rarely used because they were over land and therefore not as safe as Purple Anchor which was 20+ miles off the coast over the South China Sea.

The Boom Operator and I would fly the airplane to Vietnam and set up in Purple Anchor. He would wake the A/C and the Nav when the first chicks came up. When the A/C returned to the left seat, he would pull on his helmet and always say "Where are they." I would answer, "40 miles, five o'clock" or whatever their position was.

Most refueling was done at 16,000 feet. Often there would be several tankers on station, so they were "stacked" at 4,000 foot intervals. When the war was slow and the chicks were few and far between, there might be tankers at 16,000, 20,000, 24,000, and maybe even 28,000 feet. When the bottom tanker reached bingo fuel, he would depart the anchor and the remaining aircraft would descend to the next lower altitude.

If chicks were already on the tanker at 16,000 and new chicks arrived, they would go to the tanker at 20,000.

F-4 MAYDAY IN PURPLE ANCHOR

One afternoon we were in Purple Anchor, not the low tanker, probably at 20,000 feet waiting our turn. All aircraft monitored a "GUARD" frequency that was used for emergency transmissions. On the VHF radio it was 121.5 and on the UHF it was 243.0. That afternoon an aircraft came up on GUARD with "Mayday. Mayday. Mayday." He explained that he had received some sort of battle damage and that although his airplane was flying, he was losing fuel at a high rate. The A/C, being an old fighter pilot himself, jumped at the chance to help the guy. We gave him a radio frequency to use and we both switched over. After a brief conversation, we learned that he was a Fast FAC (Forward Air Controller) F-4 coming out of Laos so we turned out of Purple Anchor toward the coast. The Nav gave him an IFF (Identification Friend or Foe) squawk code so we could positively identify him on our radar, and we made a course straight at him. We made a typical rendezvous where the tanker and receiver fly directly at each other and on the Nav's direction the tanker does a 180 degree turn in front of the receiver. He got into position for refueling and our Boom Operator flew the boom into position and extended it out to his fuel receptacle.

The latches on his receptacle would not allow the boom to fully enter and latch. We held down-pressure on him, and he held up-pressure on us. I turned on all six refueling pumps when, for an F-4, you would normally only use two. Fuel was streaming back over the top of his airplane but enough was getting into his system to just replace the fuel he was burning. We "dragged" him about 30 minutes down to Da Nang and turned him loose right over the base so that he could land with a minimal amount of fuel.

We broke a whole list of rules that day, leaving Purple Anchor, getting too close to the coast, holding down-pressure on the receiver, etc. When we left to return to CCK, the A/C said, "This did not happen. Do not tell anyone what we did today." I thought he was being melodramatic.

Many years after the war was over, I came across a story of a similar, albeit much more dramatic and critical, incident where a tanker over Thailand had gone north to rescue an F-4 that had only one good engine and was badly shot up. When the tanker crew returned to base they were met with a threat of court-martial. The only thing that saved them was that the F-4 unit put the tanker crew in for a Silver Star. The result was that nothing was done. In light of that story, I have to admit that our A/C was right to keep our story to ourselves.

I have always wondered how many other similar stories went untold.

HOME EARLY

One night in November of 1972 on our way to Vietnam, the Boom Operator and I were listening to Radio Hong Kong on the HF radio. We heard Henry Kissinger say, "Peace is at Hand." We didn't know what that would mean for us, but we were not expecting anything in the near future.

Under the terms of the peace agreement, the United States could maintain in South Vietnam whatever level of support existed as of January 1, 1973. In order to maximize South Vietnam's ability to defend itself, an effort was made to transfer as much military hardware as possible to South Vietnam before that deadline. The South Vietnamese Air Force flew F-5's, which were the fighter version of the Northrup T-38, our supersonic trainer used in undergraduate pilot training. The USAF did not operate the F-5 fighter version, so did not have any to transfer to South Vietnam on such short notice.

The Republic of China (Taiwan) Air Force *did* operate F-5's. An agreement was made that Taiwan would give their F-5's to the US so they could be transferred to South Vietnam before the January 1, 1973 deadline. The condition was that the USAF had to send F-4's to

Taiwan to secure their airspace until replacement F-5's could be manufactured.

The base chosen as a temporary home for the USAF F-4's was CCK. There was not room there for both the F-4 operation and our tanker operation, so we were sent home to Barksdale. The other four Provisional Squadrons of tankers that had deployed in May 1972 like we had, all stayed in place until March 1973 or so. We got to come home on December 4, 1972, in time for Christmas.

Our crew flew as passengers on the leg from CCK to Hickam in Hawaii. From Hickam to Barksdale, we were the flight crew. Between Hawaii and the west coast there were three routes; we were to fly the most southerly route. Across the open water we had a quartering tailwind of about 125 knots. When we made a slight turn north to "coast in" around Los Angeles, that tailwind was directly behind us.

Most airplanes of the size and type of a KC-135 would cruise at a power setting that optimized fuel consumption. This was slightly slower than a maximum speed but it conserved fuel. Being a tanker, we had no such concern; fuel for our use and fuel to offload to other airplanes was interchangeable so fuel consumption was not a problem for us. We had a solid airplane and lots of fuel, so we flew fast. When we coasted in with that 125-knot tailwind we were doing 650-675 knots ground speed. I was the only one of the crew returning to a wife and family. The others were returning to empty apartments right before Christmas. The A/C said, "You don't even need to look at the airspeed indicator to know how fast we are going, just check out Gravel's smile."

As we approached west Texas, we encountered a line of thunderstorms that stretched from Mexico to Canada. They were high and still building. We kept requesting higher and higher altitudes and finally slipped through a gap between two storms over Dallas/Fort Worth at around 43,000 feet. The lightning show was amazing. I put my camera in the windscreen and just held the shutter open for some random length of time. I got one good picture that Sheri later framed. It hung in the kitchen for years.

BARKSDALE

FOOD STORIES

From where we lived in Princeton, a few miles east of Barksdale AFB, I would pass a Southern Maid donut shop on my way to the base. One day I was going in for flight planning and stopped for a few donuts to eat with my coffee. The Nav asked me if he could buy one. The next time I just bought a dozen and easily sold the ones I didn't eat. We started "bootlegging" donuts. Ultimately, I would buy six dozen or so donuts and the Nav would sell them to the other crews. We ate donuts for free from the proceeds of our sales.

When Sheri and I lived in Austin, TX, for graduate school, there was a chain of steakburger restaurants there called the NightHawke Restaurants. It was pricey for a hamburger joint, but the food was delicious. One day when I was in the Commissary at Barksdale, I noticed that the Nighthawke chain was selling a TV dinner, so I tried it and as TV dinners go, it was top-of-the-line.

Because we were trying to burn as much fuel as we could in those days, we flew very long training missions. We had one mission flight-planned that we would fly clockwise one time and counter-clockwise the next. Leaving Barksdale we flew to the Dallas/Fort Worth area where we would rendezvous with a B-52 for re-fueling, then continue to the southwest, turn north over Lake Tahoe, turn east over Montana, skirt just south of Chicago, turn south over the Carolinas, do a U-turn over central Florida, and then west over Georgia to recover at Barksdale. When we flew it counter-clockwise we might refuel the B-52 near Tampa.

Because these were long missions, we had to take a meal with us. The In-flight Kitchen offered a cold sandwich in a box with chips, fruit, etc. The other option was the Air Force version of a TV dinner. We had a small oven onboard and the TV dinner was just the right dimensions to fit into the oven.

I don't remember how, but at some point it occurred to me that the Nighthawke TV dinner was the same size and shape as the one from the In-flight Kitchen. We tried it and it worked. From that point on, we would bring the Nighthawke dinners bought with profits from donut sales. Whenever we had a check pilot or Wing staff or other visitors on board, they always envied our fare.

40 ATTA-BOYS

After we returned to Barksdale, we flew a lot of training missions because we needed to burn our fuel allocation that had not been burned while we were in Southeast Asia so that we wouldn't lose it the next year. A decision was made that co-pilots with less than 18 months left in the service would not be upgraded to aircraft commander. I had 17 months left. After a short time, I was so senior as a co-pilot that they didn't really know what to do with me. The solution was to create a new position and make me an Instructor co-pilot. This meant that my crew would always have two co-pilots, me and the new guy right out of pilot training. One good result was that I rarely had to sit alert since the younger co-pilot needed the experience more than I did.

This left me with lots of time on my hands. One day during flight planning I realized that the end-of-runway obstruction that we used for flight planning for our training missions was different from the one that was used for our Emergency War Order (EWO) flight planning, the nuclear war scenario. After some investigation, I realized that the EWO obstruction was outdated and incorrect and therefore the nuclear war planning, which was the reason for our existence, was badly flawed.

I told the A/C about this and he told me to take it to the Squadron Commander. When I did, he wanted me to go with him to the Wing Commander. When I presented the information, the Wing Commander thanked me for my discovery and promised to get the problem corrected. He was likely glad that I had discovered this instead of it being discovered during an Inspector General inspection or something like that.

We left the Wing Commander's office and were walking on the sidewalk back to the Squadron. The Squadron Commander said, "Son, you did good here but just remember, it takes 40 Atta-Boys to make up for one Oh, Shit."

FIRST TEAM BRIEFING

After the end-of-runway obstacle incident, the Squadron Commander seemed to take more notice of me. So many pilots were leaving the Air Force at that time, I think they might have been trying to change some of our minds. One day I was told that they wanted me to fly to Offut AFB, the headquarters of SAC, for something called First Team Briefings. This was described as a real honor to be selected and I didn't have anything else pressing so I willingly went along.

At the time, I was a young Captain and about as low on the totem pole as anyone at the Briefing. They talked a lot about the capabilities of SAC and what the future held and portrayed the career opportunities very positively. I kept thinking about those poor B-52 crews we had gotten to know when we were on Red Alert in Guam.

One thing that they said really got my attention and stuck with me until today. They BRAGGED that during the bombing of North Vietnam, they had real-time control of the bombing. They had radio contact with the airplanes and could say yes or no to each push of the button. My reaction to that was that I would never want to work for an organization that wanted to micro-manage me like that. They would want me to take all the risks involved in getting over the target and then change their mind about how important it was. I don't think so! I kept my mouth shut but the trip did not have the effect on me that they were going for.

While at Clark in 1972, we had a young staff attorney who had been sent from Offut. I don't know what he did that summer, but we had gotten acquainted and struck up a friendship. Before going to Offut for the First Team Briefing, I contacted him to let him know I was coming. He insisted that I stay with he and his wife at their base housing home. We had a great visit and we regaled his wife with stories of all the goings-on at Clark that summer, the "Bullshit Disapproved" stamp, and all of it.

At their home for breakfast was the first time I ever ate an English muffin. A southern boy from a poor background discovered a new dimension to breakfast. To this day, an English muffin with butter and orange marmalade is one of my favorite breakfast sweets.

EPILOGUE

As I approached my Date of Separation from the Air Force, we had to make a decision about where we would move and how I would make a living. I wanted to get back into Engineering. I wanted my kids to know my siblings and their cousins. I wanted to live in a place with economic opportunity so that I could provide a comfortable life for our family. My older sister and brother were in the southeast, so I focused on that area. I made a list of cities which might be suitable. It didn't take long listing the pros and cons for Atlanta to become the obvious choice.

It was reasonably close to my siblings in Huntsville and Oak Ridge. It had multiple civil engineering consulting firms in case I chose the wrong one initially. It was in Georgia where my Caribou buddy David Mitchell was from. And, for the icing on the cake, the Air Force Reserve unit at Dobbins ARB in Marietta was flying C-7A Caribous.

I went to the Shreveport public library to find a Yellow Pages Directory for Atlanta. I wrote down the names of all of the Civil Engineering Consulting firms that were in bold print. I sent them each a resume and followed up with a phone call a few days later. Every firm asked that I come by for an interview when I came to Atlanta. What I didn't know at first was that the Federal Government was promising a huge increase in funding for rehabilitation of sewer systems and all engineering firms that did that type work were gearing up to get some of that money.

My last week in the Air Force I took some accumulated leave and drove to Atlanta. Upon arrival in the afternoon, I started calling the firms I had contacted. The first was Black, Crow and Eidsness. They were based in Florida but had recently opened an office in Atlanta and were hiring. I made an appointment for the next morning. I arrived at their office about 9:00 am and was introduced to the boss. He was very gracious, spending several hours with me and taking me to lunch. After lunch we returned to his office and he told me that although he could not offer me a job, he could assure me that they would be offering me a job and that the salary would be about $11,500. That was a couple thousand more than I was expecting so this was a big boost to my spirits. Then he asked what other firms I was interviewing with and I told him. He said, "You've got all the right ones, except you might consider talking to Jordan,

Jones, and Goulding." I didn't take much note of this because in my mind I was on the verge of accepting his offer as soon as it came.

The next two days I interviewed three or four other firms and remained convinced that my first interview was the best by far. The day I was going to leave Atlanta to head home, I completed a morning interview downtown and was headed up I-85 to my motel to pick up my things and leave. As I drove, the comment about Jordan, Jones & Goulding kept bouncing around in my head. I couldn't make it go away. I asked myself why their competitor would have said that if there was not something about that firm that he admired. I decided to give them a call and see where it would lead.

I took the next exit which was North Druid Hills and quickly found a service station that had a pay phone with a phone book. I found JJ&G easily enough and called them. The receptionist transferred the call to Mr. Jordan, and he encouraged me to come by. In a crazy coincidence that I took as a sign, their office was just south of North Druid Hills, right on the other side of I-85. It couldn't have been more than a mile from the phone booth where I was standing.

The minute I walked in the door of the office the place just felt like home. I talked with Mr. Jordan and a few others for a while that afternoon. They asked if I could come back the next morning to talk to Mr. Jones. I did and he offered me a job. They had an appreciation and respect for my military service and my aviation training. They were very nice people. They offered me a comparable salary and the office was not downtown.

All the way back to Shreveport, I agonized about how I would call the boss at BC&E and tell him that I had taken his advice and was accepting a job at his competitor. He had been so nice to me that I just couldn't see how I could do that. Then I realized that I was allowing a few uncomfortable minutes on the phone to cause me to do something that I would regret for a long time. Once at home and settled down, I made that call and he was very gracious about it.

My direct supervisor for the seven years I worked at JJ&G was Jimmy Williams. Jimmy became a surrogate father to me, giving me good advice and teaching me so much it is hard to imagine how my career would have progressed without him.

One year later, our son Alan was diagnosed with leukemia. JJ&G had a 100% major medical insurance program which relieved me of all financial burden which would have exacerbated the tragic circumstance of his illness.

When Alan was diagnosed, Emory Clinic and Egleston Children's Hospital had just hired a new Pediatric Oncologist and Alan became his first patient in Atlanta. We could not have been in a better place for Alan to receive the absolute best treatment under the most positive circumstances. When we left the USAF a year earlier, we could not have anticipated this turn of events, but it was our good fortune to be in the right place to deal with it. Alan fought his leukemia bravely for five years and had as normal a life as possible for the first 4-1/2 years. The last months were not so normal and he died at Egleston Children's Hospital on April 19, 1980, a few months after his 10th birthday.

I never flew Caribous in the Reserves. When I arrived in Atlanta, they did not have any slots open. Later when they did, Sheri and I were too busy managing Alan's illness and I did not want to take that time away from my family.

The Federal funding for sewer rehabilitation did not materialize as expected. The job market for entry-level engineers in my field went from every firm hiring to many laying off and no one hiring. Eighteen months after we arrived in Atlanta, BC&E closed their new office. Had I accepted that first job offer I would have found myself unemployed in an environment when jobs in my field were virtually impossible to find.

It does not escape me that only a few of the reasons that we selected Atlanta as our home turned out to be relevant and most were not relevant at all. Looking back, however, it is hard to imagine having chosen any other place to live. In this and in many other ways, I have been blessed in my life.

In the course of my life, I have experienced many high points and triumphs but even in the tragic and difficult circumstances that I have encountered, I was blessed with the love and substantive assistance of a wonderful wife, my family and friends, a benevolent employer, and a boss who guided me like a loving father.

Sheri, Alan, and the twins Ryan and Mark.

AUTHOR'S NOTE

In 2005, I was interviewed by a small group of English students at Georgia Tech. They interviewed a number of Vietnam Veterans and asked each the same nine questions. The first eight were basically name, rank, and serial number questions. The last and key question was, "What do you consider the most satisfying factor about your service in Vietnam?"

Answering that question was easy. In both Caribous and Tankers, our role was directly supportive of our fellow servicemen who were most at risk. Caribous were a big part of the life-line for the Special Forces and forward artillery bases who were most exposed to enemy attack. We delivered their ammunition, food, mail, replacements, and a host of other goods that enabled them to take the fight to the enemy.

In Tankers, we re-fueled the F-4's and others who were directly involved in attacking the enemy and taking huge personal risks to accomplish the objectives that all of us were there for. My sense of duty was to those front-line soldiers and airmen and my greatest satisfaction was knowing that I did my best in that effort.

Despite all of the negative feelings and attitudes associated with the Vietnam War, over 90% of Vietnam Veterans would serve again if asked. I am proud to say that you can count me among them.

ABOUT THE AUTHOR

Alan Gravel grew up in Alexandria, LA, with three siblings, two older and one younger. When Alan was ten, his father died, and the family survived by the sheer determination of their mother. After attending Bolton High School, Alan earned a BS in Civil Engineering at Louisiana Tech and then an MS in Environmental Health Engineering at the University of Texas. He married the late Sharon (Sheri) Aasen in 1968 and in early 1969 joined the USAF. After Officer Training and Pilot Training, Alan flew C-7A Caribous in Vietnam. His next assignment was KC-135 Stratotankers at Barksdale AFB. In tankers, he served two temporary duty tours in Southeast Asia. After leaving the Air Force in 1974, Alan and Sheri moved to Atlanta, GA, where he now resides. Alan and Sheri had three boys: Alan Watson, who was born during pilot training and died in 1980 of leukemia; and twins Ryan and Mark, who live and work in Atlanta. Upon return to civilian life, Alan joined a Civil Engineering consulting firm in Atlanta. In 1981, he accepted a position with a small heavy civil contractor, doing mostly water and sewer plant construction. In 1992, he formed Willow Construction. He now lives in Marietta with his wife April.

CPSIA information can be obtained
at www.ICGtesting.com
Printed in the USA
LVHW080347180222
711021LV00001B/1